WASTE MANAGEMENT

Waste Management

Towards a sustainable society

O.P. Kharbanda
and
E.A. Stallworthy

Auburn House
New York · Westport, CT · London

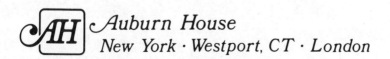

Library of Congress Cataloging-in-Publication Data
Kharbanda, Om Prakash.
Waste management.
Includes bibliographical references.
1. Refuse and refuse disposal. I. Stallworthy, E. A. (Ernest A.). II. Title.
HD4482.K47 1990 363.72'8 89–18496
ISBN 0–86569–000–6 (alk. paper)

Library of Congress Catalog Card Number: 89–18496
ISBN: 0-86569-000-6

First published in 1990

Auburn House, 88 Post Road West, Westport, CT 06881
An imprint of Greenwood Publishing Group, Inc.

10 9 8 7 6 5 4 3 2 1

Contents

PART IV THE ENVIRONMENTAL ISSUES

PART V THE ROLE OF MANAGEMENT

Tables

Figures

Abbreviations

ACGIH	American Conference of Government Industrial Hygienists
AEA	Atomic Energy Authority
BPEO	Basic Practicable Environmental Option
CEFIC	Conseil Européen des Fédérations de l'Industrie Chimique
CFC	chlorofluorocarbon
CMC	Chemical Manufacturers' Association
DEC	Department of Environmental Conservation (New York)
EC	European Commission
EEC	European Economic Community
EPA	Environmental Protection Agency (US)
FGD	flue gas desulphurization
GWP	gross world production
ICI	Imperial Chemical Industries
ILO	International Labour Organization
IPC	Indian Plastics and Chemicals
IPM	integrated pest management
LPHC	low probability, high consequence
MCA	maximum credible accident
MVP	multivalent phenol
NAPAP	National Acid Rain Precipitation Assessment Program
NCRR	National Centre for Resource Recovery (US)
NIMBY	'not in my back yard'
NLC	Neyveli Lignite Corporation
NRDC	National Resource Development Corporation (US)
OCS	Ocean Combustion Service BV (Rotterdam)
OECD	Organisation for Economic Cooperation and Development
OTA	(US Congress) Office of Technology Assessment
PCB	polychlorinated biphenyl
PNC	Power Reactor and Nuclear Fuels Corporation (Japan)

POWER	Pipeline Operated Waste Energy Repository
RCRA	Resource Conservation and Recovery Act, 1976 (US)
RDF	refuse-driven fuel
SFR	a Swedish under-seabed method of storing high-level radioactive waste
TSCA	Toxic Substances Control Act (US)
UN	United Nations
UNECE	United Nations Economic Commission for Europe
UNEP	United Nations Environmental Planning
WHO	World Health Organization
ZID	zero–infinity dilemma

Preface

The problem of waste management is rapidly assuming enormous proportions, a problem intensified by our 'throwaway' society. Unfortunately, many of the solutions adopted have been largely of the 'firefighting' type, without any real regard to the long-term implications of the problem and its solution. With such an *ad hoc* approach, the real issues involved have not been carefully and clearly set out. Indeed, in the process, even more serious problems have often been created. There are many grim reminders of this right across the globe, and we present some of them as examples of what *not* to do. The problem is worldwide and whilst it is at its most urgent in the developed world, there is patently a very similar and growing threat to the developing world.

Our particular approach to this problem is not to write a scientific reference text on waste handling, but to highlight the various hazards involved in waste production and handling, and hence the dire need for proper waste *management*. We advocate a non-waste technology as the best possible approach, minimization as the next best and waste treatment as the last option, and the least desired. We realise, however, that the world has still a long way to go and the cost of the proper handling and disposal of waste is probably the most significant delaying factor.

Waste, if not properly treated and handled, not only threatens human life in the short term, but the environment as a whole in the long term. Warnings of this nature were first raised in 1962 by the missionary Rachel Carson, now seen in retrospect to have lived well ahead of her time. The warning was sounded in her book *Silent Spring*. The fears she then expressed, which were based on very limited data, have since been largely confirmed. There is even a sequel with a most appropriate title, *Silent Spring*

Revisited, published by the American Chemical Society. The point at issue here is the ever-increasing use that farmers are making of pesticides because of their immediate benefits – although lately even these are suspect. Fortunately, various concepts of farming that minimize the use of pesticides, such as integrated pest management, or even their complete elimination via organic farming, are gathering momentum. But farmers are not alone in their short-term approach to the problem: industry as a whole has as yet failed to grasp this particular 'nettle'.

Waste dumps, landfills and incineration, the most common disposal techniques in most countries, have proved to be no solution. The toxic, hazardous waste is only hidden away as such, or as ash, and surfaces after a period of years, sometimes with disastrous results. The problem is fundamentally one of proper *waste management*: waste has to be 'managed', not just disposed of. This is an immediate problem that can, and should be, tackled by management *now*, and at its roots. Existing industrial processes should be assessed and ways sought to minimize or eliminate such waste as they create. New processes should not be introduced until they have been exhaustively studied from this point of view: proper waste analysis at the design stage is imperative. Not only should the immediate situation be assessed, but also its consequences in relation to others. For instance, plastic bottles may have economic advantages for the drinks manufacturer when compared with glass bottles, but they pose a serious environmental threat. Industry should look for processes and products where there is no waste – the so-called 'non-waste technology'. Failing that, the quantity of waste, whether harmless or not, should be reduced to the barest minimum: what we haven't got cannot hurt us! Where waste is inevitable, can it be recycled for profitable use or can some valuable resource be recovered? Chemical or biological degradation must be the last resort. Incineration sounds a very acceptable solution, but it is expensive and unfortunately there is not only the problem of fumes, but the residual ash is often very dangerous, since it can be toxic.

In this book we begin, in Part I, by considering the scope and extent of the problem: society faces a staggering cost in public health, human lives and the threat of a permanently blighted environment. Because of mounting public concern, in Part II we consider the major methods of waste disposal and treatment, thereby bringing us to the crucial aspect of our subject – finding the real answer to this problem – which we discuss in Part III. Whilst the various technologies covered in this part of the book, such as non-waste technology, recycling and resource recovery, are already being practised to some extent much still remains to be done in these 'clean technology' areas. They can provide simple, innovative and economic solutions to the seemingly intractable problem presented by hazardous waste. The intractable problem of waste *must* be tackled at its 'roots'.

The impact on the environment, together with the biological treatment of waste, are covered in Part IV. The poisoning of ourselves and our environment *can* and *must* be stopped, and having outlined some typical approaches, we turn to consider, in Part V, the crucial role that management can, and should, play in these vital areas. The entire ecosystem must be the concern of management.

Waste management affects us all. We have therefore written this book with the objective of more fully informing not only those directly involved, such as health and safety officers, but also technical managers, plant managers and operating staff, central and local government officers responsible for the environment, pollution and the like: all of whom, while well aware of the issues involved, desire to be ever better informed and equipped in order that they may do their jobs well. All involved with waste need to develop sound policies in that context and we need to remember that 'policy makers' are by no means confined to governments, local authorities and the like. Policy makers in effect exist in every company, large or small, and we are proposing that one of their key policies be the *proper management* of waste.

It is indeed our hope that this book will contribute to an increasing awareness of the nature and the extent of the hazards posed by waste and the steps that can, and should, be taken to eliminate or at least minimize them. Waste will always be there to be disposed of, but let us first reduce it to the irreducible minimum and then ensure its safe disposal.

Then too, there is a very fine line between waste and useful product. For instance, the notorious PCBs are actually valuable products in their own right.

Man is far from perfect, but the systems which he designs and operates can be so organized that there is the minimum of damage not only to his fellow human beings, but to the environment in which we all have to live.

O.P. Kharbanda, Bombay, India
E.A. Stallworthy, Coventry, UK

Acknowledgements

The references listed at the end of chapters not only provide a guide for those of our readers who want to research further into any particular topic discussed: they also represent an acknowledgement of our debt to the many researchers and commentators who have written on the various aspects of waste management and control that we touch upon in the course of this book.

A book such as this is by no means the product of we two alone, even though it bears only our two names. We have learnt much from our teachers, our co-workers and our students over the years. We are also most grateful to the many authors, listed separately in the author index, who are cited as direct sources in the references. It is their hard labour that has made this book possible.

Librarians are among our benefactors, and our grateful thanks go to staff at the following libraries and institutions for their labour of love in tracking down the material we were seeking:

- In Bombay
  The American Library
  Bhabha Atomic Research Centre
  The British Council
  Industrial Credit and Investment Corporation of India
  Indian Institute of Technology
  National Institute for Training in Industrial Engineering
  University of Bombay

- In the UK
  British Institute of Management
  Institute of Bankers

London Business Library
Science Reference Library

- In the USA
  New York Public Library
  Fairleigh Dickinson University Library, Rutherford, NJ
  Library of Congress

Finally, although separated by a continent and a culture, we would both acknowledge our need for the loving support of our wives, Sudershan and Dorothy. Our thanks go out to them for their continuing patience and forbearance.

OPK
EAS

Part I
INTRODUCTION

Chapter 1

WHAT COMES FROM WHERE?

Before coming to consider 'what [waste] comes from where' let us be clear as to what waste actually is. In approaching our subject, *Waste Management*, we have first of all to decide what we are proposing to 'manage'. It seems that the term 'waste' defies precise definition, but we can make a start with our constant friend and guide, the *Concise Oxford Dictionary*. That tells us that this versatile word can be used as an adjective, a noun and a verb. For our present purpose, we are concerned with its use as a noun, but even then its meaning varies widely. It seems that our word 'waste' is partly of French origin, and that there are a number of different meanings attributed to it:

1 Desert, waste region, dreary scene: e.g., a waste of waters (unbroken expanse of sea or floods).
2 Waste material or food, useless remains, refuse, scraps, shreds.
3 Injury to estate caused by act or neglect, especially by a life-tenant.
4 Waste-basket, especially waste-paper basket: waste-pipe, to carry off waste material, especially from washing, etc.

In this present book we are chiefly concerned with the second of the definitions above. Although waste is obviously undesirable, it is nevertheless an inevitable and inherent product of our social, economic and cultural life. Will it ever be possible to advance 'beyond the age of waste': to get to a position where there is no waste? Whether or not this most desirable objective is achievable, we must most certainly strive for it. This prospect is considered vital enough to warrant a full-length book with that title: *Beyond the Age of Waste – a report to the Club of Rome.*[1] This 'club' consists of some of the most distinguished thinkers of our times, who came to the fore following the publication of its highly stimulating but controversial book *The Limits of Growth* in 1972.

This particular report covers a fairly wide canvas and examines the world situation in the face of diminishing natural resources, the energy crisis and the provision of adequate food supplies. It is the work of a group of internationally known scientists and experts in a wide variety of disciplines, who analysed the various issues both from the humanitarian and the technological aspects. The report has proved to be a very important document, in that it has stimulated a greater awareness of the serious global problems that are threatening the quality of life. It has thereby contributed to a better understanding of the need for the very drastic action that will have to be taken if a global equilibrium is to be maintained.

The information explosion

It seems that in almost every field of human endeavour there has been a virtual explosion of information. The volume of information, in physical terms, seems to be doubling every five to seven years. Our present subject is no exception to this rule. Indeed, we believe the rate of growth in this field is substantially higher than the average, since the issues involved are basic to human survival and therefore demand urgent study if appropriate solutions are to be found. Apart from the hundreds of articles that have appeared in scores of journals published worldwide, there are scores of books being published every year on various aspects of the subject. Naturally, much of the attention has been directed towards the handling, treatment and management of hazardous waste. Typical of the numerous handbooks that have appeared dealing with hazardous waste is that edited by H.M. Freeman: *Standard Handbook of Hazardous Waste Treatment and Disposal.*[2] Another, a looseleaf volume with the simple but direct title, *Hazardous Wastes Handbook*, deals mostly with the legal aspects of hazardous waste handling, the authors all being lawyers.[3] In view of the growing consciousness and recognition of public concern in this area, the legal aspects are indeed crucial. There are some 300 million tonnes of hazardous waste produced each year in the United States alone and this needs to be very strictly controlled if the safety of the general public is not to be imperilled. The purpose of the handbook is: 'To help readers gain a working knowledge of the massive hazardous waste management program'. It is of course completely oriented to the United States and its legal system.

To give some impression of the vast storehouse of literature available for study, we list below a few of the titles that have been published over the past few years:

Allegri, T.M. (1986), *Handling and Management of Hazardous Materials and Waste*, USA: Chapman and Hall.

Conway, R.A. and Ross, R.D. (1980), *Handbook of Industrial Waste Disposal*, New York: Van Nostrand.

Davis, C.E. and Lester, J.P. (eds) (1988), *Dimensions of Hazardous Waste: Politics and Policy*, London: Greenwood Press.

Dawson, G.W. and Mercer, B.W. (1986), *Hazardous Waste Management*, USA: Wiley Interscience.

Fawcett, H.H. (1984) *Hazardous and Toxic Materials – Safe Handling and Disposal*, USA: Wiley Interscience.

Holmes, J.R. (ed.) (1983), *Practical Waste Management* Chichester: Wiley.

Martin, E.J. and Johnson, J.H. 3Jr. (1987), *Hazardous Waste Management Engineering*, New York: Van Nostrand.

Pierce, J.J. and Vesiland, P.A. (1981), *Hazardous Waste Management*, USA: Ann Arbor Science.

Robinson, W.D. (1986), *The Solid Waste Handbook – A Practical Guide*, USA: Wiley.

Sittig, M (1981), *Pesticide Manufacturing and Toxic Material Control Encyclopedia*, USA: Noyes Data.

Sweeney, T.L., Bhatt, H.G., Sykes, R.M. and Sprou, O.J. (1982), *Hazardous Waste Management for the 80's*, USA: Ann Arbor Science.

Voight, R.L. (1986), *Waste Management and Resource Recovery*, USA: International Resource Evaluation.

Woolfe, J. (ed.) (1981), *Waste Management*, USA: Kluwer.

The above are perhaps typical of the scores of such books that are being published every year. We must, however, make it clear that the books we have listed are merely a random selection; nevertheless, their titles give a 'flavour' of the available knowledge on the subject of the handling and management of waste.

Waste management is a crucial aspect of waste handling, particularly when we come to consider its *safe* handling and, in *Safety in the Chemical Industry*, we ourselves have contributed to this particular aspect of the subject.[4] In a section headed 'Storing up trouble' we outline the problems that were encountered following the burial of chemical waste in a place called the 'Love Canal', near Niagara Falls. We found the placename 'Love Canal' intriguing, and wondered whether those who visited the Falls were wont to take an idyllic stroll along its banks and gaze together into its clear waters. But no! The canal was named after a certain William T. Love who came to Niagara Falls in 1892 and planned to build a navigable canal there. The project collapsed and the canal – or rather, ditch – became the property of a chemical manufacturing company who used it for the disposal of hazardous chemical waste. This buried waste was later disturbed, with sad consequences for those who lived in the neighbourhood. We cannot retell the story here, but the case made front-page news and resulted in media attention being drawn to the multitude of very similar unsafe waste dumps scattered across the USA. A comprehensive cover story in the *National Geographic Magazine* in this context illustrates very well the type of work and the progress being made in relation to waste disposal and handling in many parts of the world,[5] and we recommend its careful perusal. Another source of valuable background data are the annual volumes issued under the title 'State of the world' as part of *A Worldwide Institute Report on Progress Toward a Sustainable Society*. The 1987 volume[6] contains a 21-page chapter titled 'Realising recycle potential', whilst the 1988 volume carries a 19-page chapter on 'Controlling toxic chemicals'.[7] The latter is a condensed version of Worldwatch Paper No. 79: 'Defusing the toxic threat – controlling pesticides and industrial waste'. This particular 'annual volumes' project is under the direction of the indomitable Lester R. Brown.

What and how much waste from where?

It is obvious that, for proper waste management, sound answers must first be established to the following three leading questions:

1 What is the nature of the waste?
2 How much is there of it?
3 Where does it come from?

Unless we have a clear picture as to the type and volume of waste that comes from the various sources, we cannot possibly plan its proper management, which must include its handling, treatment and safe disposal. These several aspects of waste management are of course discussed in detail in later chapters, but we wish initially to 'take stock' of the situation and set up what we might call an 'inventory' of the various types of waste that have to be dealt with in the normal course.

When we come to consider waste, we find it an area where hard data (facts and figures) are very difficult to ascertain. But one thing is certain. Waste management is undoubtedly a growing and very profitable business. The infant waste management industry is estimated to be now growing at the rate of some 20 per cent per year.[8] But there are numerous problems. For example, the relevant legislation is constantly changing and becoming ever stricter. While discussing this aspect in relation to hazardous waste management, Geraldine V. Cox, the vice-president and technical director of the Chemical Manufacturers' Association, has observed:

> If someone would tell us what to do once and let us do it, we would be happy. But they keep changing the goals as we're going through the process, which makes compliance very costly and not necessarily very productive.[9]

The various changes and amendments that are made to the regulations are usually either the result of new information or an increased public perception of the risks involved. Neither of these can be foreseen.

The three main sources of waste

Broadly speaking, the great bulk of the waste that has to be handled and disposed of is generated by three major sources: domestic refuse, agricultural waste and industrial waste. The type of waste from domestic and agricultural sources is generally similar, irrespective of the area or even the country where it arises, but industrial waste varies very widely, being largely dependent upon the type of industry involved. Some idea of the total volume of domestic waste that has to be handled worldwide can be gauged from the fact that the volume to be disposed of from New York City is said to be of the order of 30,000 tonnes per day. This domestic waste contains not only valuable and often reusable materials such as metals, glass, paper, plastic and food waste with a high nutrient content, but also an ever-increasing amount of hazardous waste. Typical of the latter is mercury from batteries, cadmium from fluorescent tubes, pesticides, bleaches, PCBs in old TV sets and a wide range of toxic chemicals such as occur in solvents, paints, disinfectants and wood preservatives.[10]

INTRODUCTION

Some facts and figures in relation to this 'garbage glut' make sorry reading. Whilst the per capita generation of domestic waste is currently estimated at around 2 kgs per day in New York City, it is said to be half that much in Hamburg, Hong Kong and Singapore, and only about a third as much in Sandung (Indonesia), Calcutta (India), Kano (Nigeria), Manila (Philippines), Medellin (Colombia), Rome (Italy) and Tunis (Tunisia). One unexpected feature of the review which produced this data was the fact that the per capita quantity of domestic waste does not seem to vary much with the state of development of the country concerned. As will be seen from the listing of cities above, some are highly industrialized and others are still developing, yet they have comparable rates of generation of domestic waste. Nevertheless these are huge quantities and it would seem that the priority must be to minimize its initial production and recycle as much as possible. In other words, we must tackle the problem at its roots.

When we turn to consider the agricultural industry, we see loss and waste occurring in this industry's products at every stage in the long chain from farmer to consumer. In terms of volume, the major losses occur at harvesting (about 20 per cent), during processing (10 per cent) and during usage (10 per cent). The loss on usage includes that at the dinner table, where improper distribution of what is available amongst the various members of the family can result in food being thrown away rather than eaten. Further substantial wastage can also occur due to the poor absorption of food in the human body, as a result of disease. It is clear that, in seeking to solve the problems created by waste in the agrosystem, much more can be achieved by striving to cut down such losses, rather than by improving the processes of production. This comment is equally valid in relation to both domestic refuse and industrial waste. At the same time, the socioeconomic structure would also be considerably improved.

It is industrial waste that poses by far the biggest problem, not so much in terms of quantity, but because of its nature and the vast variety of materials that have to be handled. Further, much industrial waste consists of hazardous chemicals, often highly toxic, that must be treated properly to ensure that they do not damage us or the environment. Ground water is especially vulnerable, and the infiltration of chemicals into ground water, which is eventually used for drinking, can be disastrous.

The quantities are suspect

It seems that there are no really reliable figures for the quantities of industrial waste that have to be dealt with, even for the developed countries.[11] The work involved in collecting such information would be very considerable and no one seems prepared to spend the effort involved. We are therefore left with a range of estimates. The total quantity of hazardous waste generated by US industry is estimated at some 300 million tonnes annually, of which by far the most is processed on-site (perhaps 96 per cent), leaving a mere 4 per cent to be handled off-site, at relatively high cost. In terms of processing costs, on-site waste costs around US$15 billion per year, whilst the off-site activity is valued at some US$2.5 billion per year. The latter is estimated to increase to some US$6–7 billion by 1991. Most – more than 90 per cent – of this total quantity of hazardous waste is in the form of liquid effluent and a mere 10 per cent is in the form of solids. The chemical industry accounts for nearly two-thirds of the total volume of hazardous waste.

Another, and independent, source of data – an article in the *Asian Review* – suggests that the per capita production of waste in the United States is about 3 tonnes per year, the comparable figure for the UK being 1 tonne per year.[12] But what does the future hold? The waste now being produced annually is expected to double by the end of this century. It seems the world is heading for a crisis, but in this crisis also lies an enormous opportunity. Such immense quantities of waste, if recycled and disposed of properly, can be of substantial help in the conservation of valuable resources. At the same time a vast quantity of energy could be generated, another aspect of waste management that we shall be discussing later. The point is that the potential is there, waiting to be taken advantage of.

Apart from the three main sources of waste that we have been discussing, there are other small, but nevertheless significant, sources. Typical of these are laboratories and hospitals. Research laboratories, whatever their purpose, can create extremely hazardous waste. One writer draws attention, by way of example, to the University of Minnesota and the North Carolina State University, where clean-up facilities have had to be provided.[13] The quantities may be small, but they can still be very dangerous, and the appropriate regulations still apply. However, there can be a tendency to hide the waste away until there is a need for a compulsory clean-up, such as when a VIP inspection looms up.

Then it may well 'cost more to burn than to buy'.[14] Much will depend upon the type of waste involved, but it can easily cost from US$250 to US$2,500 to safely dispose of a small drum (say 20 gallons) of such materials.

Hospitals are also becoming serious creators of waste, particularly hazardous waste. The refuse generated during a routine thoracic operation has been said to fill three large (30-gallon size) plastic sacks.[15] In the 1960s a similar operation would have generated little more than a handful of refuse. The exponential growth of waste in this context is largely due to the ever-increasing use of disposable items, such as paper gowns and drapes, plastic syringes and bottles, rubber gloves and catheters and other tubing, and metal needles, staples and cartridges.

Waste and the environment

There has been a growing realization that the environment has a limited assimilative and carrying capacity. This means that pollution control is essential in order to safeguard the environment and hence the quality of human life. Fortunately, there is a distinct upsurge of interest in preserving a natural and ecologically balanced environment. The problems encountered in maintaining a balanced state of the atmosphere despite the continuous emissions that occur has been clearly outlined in an English translation of an excellent German treatise on the subject.[16] Our environment must be seen as the natural capital on which we depend to satisfy our needs. Wise waste management, therefore, demands positive and realistic planning.[17] For instance, extensive reforestation is seen as a sensible way to mitigate many of the environmental problems and this will simultaneously bring many other benefits.

So far as the EEC is concerned, its environmental policy is an ambitious, long-term venture which is likely to have considerable impact in the long run. Their measures so far have been confined to 'directives'. These are not normally enforceable, except at the discretion of the member states. A directive was issued on 4 May 1976 that dealt with pollution caused by certain dangerous substances discharged into the aquatic environment of the community, but this required detailed implementation by the member states. Later (24 November 1987) a 2nd International Conference was held on the protection of the North Sea, which was attended by the Environmental Ministers from all the countries bordering the North Sea or discharging effluent into it. General agreement was reached on reducing and in some cases eliminating the dumping of pollutants into the North Sea, which has now been accorded the status of 'Special Protected Area' under the International Maritime Commission. This prohibits the dumping of rubbish in that sea by ships. Other major features of the agreement then reached were:

- Dumping of all hazardous waste banned as from the end of 1989.
- There was to be strict control on the discharge of dangerous substances into rivers ultimately discharging into the North Sea.
- Incineration of toxic waste at sea would be 'greatly reduced' by 1991 and ended by 1994. (We deal with this method of toxic waste disposal in detail in Chapter 3.)
- Efforts should be made to reduce the input of nutrients carrying nitrogen and phosphorous, which have harmful environmental consequences.
- The best available technology should be employed in the handling of radioactive discharges.

All this demonstrates that determined action is being taken to both minimize and eliminate pollution. The several EC governments have declared that there is an urgent need for technologies and practices that are more energy and material resource efficient. This would substantially decrease the waste releases into the environment and is the positive approach. But positive steps in this direction are slow in coming. Some of the targets that have been set will be hard to meet: the elimination of incineration at sea demands additional burning capacity on land and that may be hard to achieve. What is more, there is a serious doubt whether land-based incineration can meet the needs for the destruction of highly chlorinated waste.[18]

If waste is not handled properly it will poison the environment, including the air we breathe and the water we drink. According to one estimate the citizens of the United States spent more on food packaging annually than the entire revenue received by farmers in a year. Somewhere between a half and two million poisonings by pesticides are reported worldwide each year, largely among farmers in the developing world. Further, with the rapid and constant evolution of new and resistant strains of crop pests, the pesticide manufacturers are continuously developing ever more potent strains of pesticides and are locked in a race which they cannot win.

Whilst it is generally recognized that the chemicals in waste, if not properly treated and managed, pose a very serious health hazard, the problem is further accentuated by the fact that new chemicals are continually being introduced. New synthetic organic chemicals, particularly pesticides, have been brought on to the market so rapidly over the past few years that there is an utter lack of knowledge concerning their harmful effects. To take but one extreme, but nevertheless true, example, methyl isocyanate (MIC), the chemical used in the manufacture of the pesticide carbaryl, was released in large volume at Bhopal and was the cause of that dreadful and worst ever industrial disaster.[19] Yet, such was the ignorance concerning the effects of this particular chemical on the human body, that the local Union Carbide management, even after seeing dead bodies piled in heaps, declared: 'MIC is an acute irritant, but certainly not lethal'. Across the world at their headquarters in Danbury, Connecticut in the USA, their director of health, safety and environmental affairs dismissed MIC as 'nothing more than a potent tear gas'.[20] From amongst more than 50,000 chemicals listed by the Environmental Protection Agency (EPA) in the United States as hazardous, information about their toxic effects is available for only some 20 per cent. Further, a mere 10 per cent have been tested for chronic, reproductive or mutagenic effects. It seems that the indiscriminate use of pesticides without full knowledge of their effects on the human system has led to many an unpleasant surprise. Although they account for only a small percentage of some 70,000 chemicals in common use, they are used worldwide, are placed in unskilled, inexperienced hands, and at the same time present a substantial potential hazard. They pose risks not only to the farmworkers in immediate contact with them, but also to the public at large through spraying, contamination of drinking water and through being left as a residue on food crops. In fact, this is a vicious circle: a chain of events that may well have disastrous results. To quote:

> ...the more we manipulate our environment, the more we have to manipulate it. The more we use synthetic pesticides, the less we are able to do without them.[21]

The pesticide 'treadmill'

The above quotation introduces us to what we can well call the 'pesticide treadmill'. It was in her hard-hitting book *Silent Spring* that the indomitable Rachel Carson first warned us that the indiscriminate use of pesticides may well lead to serious diseases and even death. She painted a graphic picture of a world when spring would no longer be heralded by the singing of the birds. Her warning led to extensive efforts worldwide to regulate and control the use of pesticides, but it seems that these efforts have not borne any real fruit. This is evidenced by the proceedings of a seminar instituted by the American Chemical Society with the title 'Silent Spring Revisited'.[22] Pesticides are used by farmers for their immediate beneficial effects, and they may not be aware of their long-term adverse effects.

Mexico pioneered the so-called Green Revolution through the application of new technology, but in due course this spreading change in farming techniques has threatened not only the environment but society. Not only do the waste and residues cause severe pollution problems but hundreds of peasants are being poisoned every year. These and related issues are dealt with comprehensively in an excellent book based upon the proceedings of workshops sponsored by the US Environmental Protection Agency.[23] The issues discussed include who should do what, and how the costs of such actions should be shared. The damage has been extensive but, thanks to such studies, several pesticides have been banned in the West. Unfortunately these same pesticides continue to be imported and used in the developing world.[24]

It should be realized that some people defend the continuing use of pesticides. Alleged to be politically motivated, although presented as a scientific guide, Elizabeth Whalen, executive director of the American Council on Science and Health, has announced a pesticide residue book with the title *Pesticides: Do we really need them?* with the implied and emphatic answer: Yes![25] On the other hand, we can mention three articles in the popular financial press that highlight the 'perils' of pesticides. One reports on a study by the EPA which found that a chlordane pesticide widely used in homes to control termites could pose a high risk of cancer even when properly used.[26] Despite the fact that its use has been consequently banned as a termite killer, except when injected into the ground beneath homes, critics feel that this is not sufficient and call for an immediate total ban.[27] Whilst the manufacturer maintains that the use of chlordane has been thoroughly researched worldwide, one critic, the toxicologist Diane Baxter, declares:

The information we have on this stuff frightens me so much I cannot

> understand why it is still on the market ... I don't think EPA has done all they ought to do.

The third article in this vein deals with the risks involved in the consumption of produce imported from countries where pesticides banned in the United States are used.[28] It seems that, after four decades during which pesticides have been used widely, the pests are not only thriving, but new and more resistant species have come into being. The effects seem to be largely incalculable. For instance, the tobacco bugworm has become a major pest because its natural predators have been killed off by pesticides. It seems that a radically different approach is required: what amounts to a benign but effective pest control strategy, which considers the other alternatives that may be open for use apart from the use of pesticides.

Some steps have been taken in this direction – namely, IPM (integrated pest management). The approach seeks to balance the costs, the risks and the benefits of the various alternatives that may be open for use. Typical of the approaches that avoid the use of pesticides are:

- releasing sterile male insects into the affected area;
- trapping pests with chemical scents;
- releasing bacterial, oral or fungus disease;
- developing crop varieties that have a superior resistance.

Other ingenious solutions include the protection of the cotton crop by planting alfalfa alongside as a 'decoy' crop and planting late so that the insects have bred and matured well before the crop appears.

The IPM principle works not only for farmers, but also for the proper management of, for example, weeds in forests, urban parks and even the humble suburban garden. The technique is effective even against pests that attack livestock and people. The crux of the approach lies in knowing as much as possible about the pest and the way it lives in its environment. Based on this knowledge the scientist can determine the best point at which to intervene in its life cycle, and the best way to do that. IPM thus provides a simple solution to a complex and very serious problem. Not only are costs reduced, but there are many benefits. Our immediate concern is waste: this approach completely eliminates waste in many cases and minimizes it in others. This and other simple solutions to complex problems is a recurrent theme with us, and we devote most of Part III to this approach to waste management.

References

1 Gabor, D., Colombo, U., King, A. and Galli, R., *Beyond the Age of Waste – A report to the Club of Rome*, Pergamon, UK, 237pp.
2 Freeman, H.M. (ed.), *Standard Handbook of Hazardous Waste Treatment and Disposal*, McGraw-Hill, New York, 1988.
3 Hall, R.M.Jr., Watson, T., Davidson, J.J. and Case D.R., *Hazardous Wastes Handbook*, Government Institutes Inc., U.S. 1984.
4 Kharbanda, O.P. and Stallworthy, E.A., *Safety in the Chemical Industry*, Heinemann, London, 1988.
5 Boraiko, A.A. and Ward, F., 'Storing up trouble – hazardous waste', *National Geographic Magazine* (cover story), **16**, March 1985, pp. 308–51.
6 *A Worldwide Institute Report on Progress Toward a Sustainable Society*, W.W. Norton, USA, 1987.
7 *A Worldwide Institute Report on Progress Toward a Sustainable Society*, W.W. Norton, USA, 1988.
8 Rich, L.A., 'Hazardous waste management – new rules are changing the game', *Chemical Week* (cover story), 20 August 1986, p. 26+ (24 pages).
9 Ibid.
10 Ibid.
11 Hanson, D., 'Hazardous waste generation is hard to quantify', *Chemical and Engineering News*, **64**, 27 January 1986, p. 18.
12 Boyes, R.G.H., 'Don't waste it!', *Asian Review*, July 1987, p. 37.
13 Sanders, H.J., 'Hazardous wastes in academic labs.', *Chemical and Engineering News*, **64**, 3 February 1986, pp. 21–33.
14 'When it costs more to burn than to buy', *Chemical Business* (editorial), August 1987, p. 40.
15 Santos, G.H., 'Hospital waste and the medical–industrial complex', *New York Times* (letter to the Editor), 17 July 1988, p. 28E.
16 Breuer, G., *Air in Danger – Ecological Perspectives of the Atmosphere*, Revised English edition, Cambridge University Press, Cambridge, 1980.
17 El-Hinnawi, E. and Hashmi, M., *The State of the Environment*, Butterworth, London, 1987 (published under the UN Environment Program).
18 Hunter, D. 'The burning issue of waste in Europe', *Chemical Week*, 15 June 1988, pp. 19–20.
19 Kharbanda, O.P. and Stallworthy, E.A., 'The lesson of Bhopal', *Safety in the Chemical Industry*, Heinemann, London, 1988, ch. 9.
20 Picot, A., 'Fall-out from the Bhopal tragedy', *World Scientist*, September 1986, pp. 46–51.
21 Ehrlich, P.R., 'World population: is the battle lost?', *Reader's Digest*, February 1969, pp. 137–40.
22 Marco, G.J., Hollingworth, R.M. and Durrham, W., *Silent Spring Revisited*, American Chemical Society, Washington, DC., 1987.
23 Bridges, J.S. and Dempsey, C.R., *Pesticides Waste Disposal Technology*, Noyes Data Corporation, USA, 1988.

24 Lane, S., 'Pesticides use in Mexico', *The Ecologist*, **18 (2–3)**, 1988, pp. 82–7.
25 Rich, L.A., 'Pesticide alert stirs up debate on fresh produce', *Chemical Week*, 29 July 1988, pp. 17–8.
26 Shabecoff, P., 'Termite killer is cancer risk', *New York Times*, 10 April 1987, p.A12.
27 Burke, W.K., 'Chlordane critics blame EPA inertia', *New York Times*, 22 April 1987, pp. 7–22.
28 Meier, B., 'Pesticides on imported produce', *Wall Street Journal*, 26 March 1987, pp. 1–26.

Chapter 2

ACID RAIN

Both physically and figuratively, acid rain has crossed many a national frontier and has thus assumed an international dimension. For instance, pollution from coal burning power stations in the UK has been alleged to cause damage to forests and lakes in Norway and West Germany.[1] Now it seems that, after several years of controversy and many arguments, Britain has to some degree accepted responsibility and has launched a major billion pound clean-up operation. Desulphurization plants are to be installed on a number of power stations to clean up the waste gases being emitted from the chimneys. The United States and Canada have signed an agreement to clean up the Great Lakes, but neither government has as yet committed any money, nor has industry been asked to participate in the operation. Many other countries share a similar problem, and who suffers will largely depend upon the direction of the prevailing winds!

The issue of environmental pollution in general and of acid rain in particular dates back some thirty years or more, when a network of measuring stations set up in Scandinavia indicated a gradually increasing acidity of rainfall. At the same time there was a significant decrease in the fish population in northern rivers and lakes. These observations were presented to an OECD meeting in 1969, which led to extensive measurement over the whole of the OECD area. Launched in 1972, this was completed in 1977, and the observations confirmed that sulphur compounds travelled long distances. Air pollutants were found to have no respect for national frontiers. This therefore called for a closely cooperative international programme of abatement. The UN Economic Commission for Europe set up a programme for monitoring and evaluation across Europe, whilst there was a bi-lateral agreement between Canada and the United States to the same end. In due course a number of countries, notably West Germany, moved from a position of indifference to one of serious concern, once serious forest damage in areas such as the Black Forest and Bavaria had been observed.[2]

What is acid rain?

The term 'acid rain' is used very broadly to describe acidified rain, snow, dust or gas with a pH level lower than 5.6. The lower the pH level, the more acidic. Distilled water is neutral, with a pH of nearly 7.0, whilst lemon juice has a pH of a little more than 2.0. Acid rain has become a matter of great concern to conservationists and environmentalists who have been quite vocal on this issue for many years. Natural rain is, in fact, slightly acidic due to a combination of natural phenomenon, including gases from erupting volcanoes and decaying organic matter which pollute the air. However, the sulphur emissions from power stations and other industrial plant, washed down by the rain, add to and markedly increase this acidity.[3] Many industrialists, and even some scientists, tend to play down the seriousness of this particular issue. Typical of the reassuring statements that are made is, for instance, the assertion that, even if acidic emission continued at its present level for the next 10 to 20 years, the environment would not noticeably suffer. It is also said that relatively little damage is now being caused by industrial activity and, furthermore, that, although some studies have ascribed forest damage in Europe and some parts of eastern North America to acid rain, the scientific evidence does not really support this: there could well be other causes.

However, whatever the truth of the matter, it seems that the governments involved are being impelled to implement acid rain control programmes, some very expensive. A joint US–Canada study has recommended that US$5 billion be spent over the next five years in order to 'demonstrate the commercial feasibility of innovative control technologies'. These could include limestone injection, multistage burners, in-duct spraying, reburning and fluidized bed combustion. Food-grade high calcium limestone has been used successfully to neutralize and so restore 11 acidified lakes in Massachusetts and New York State. Indeed, what has been called 'liming' has been found to be a cost-effective management tool for the protection and maintenance of fisheries that are under threat from acidification. A computerized model has been developed that will determine the correct dosage of calcium carbonate in relation to the physical characteristics and chemical make-up of the water in a particular lake. The model has been so designed that it will even estimate the month and the year when retreatment will be required. Whilst such steps can go some way towards alleviating the problem, acid rain nevertheless continues to be a highly controversial issue.

A storm of controversy

Contrary to the popular conception that acid rain is a simple and well-established phenomenon, it seems to be a highly complex system. Many of the effects commonly and popularly attributed to acid rain may well be due to other factors. The *Acid Rain Report*, prepared by NAPAP (National Acid Precipitation Assessment Program) at the behest of the US government, was issued in September 1987 and immediately raised a storm of controversy. Not only the daily and financial press but also a number of scientific journals took issue with the report's recommendations and conclusions. For instance, the statement that the US$94 billion clean-up programme being advocated was described not as a cost but the 'savings consumers and taxpayers may realize' was strongly challenged in the press.[5] In any event, no government is likely to approve such a very expensive programme, particularly when the report made quite a strong case against the notion that air pollutants from coal-burning power stations were destroying the environment.

Of course, there were scientists that took the opposing view, offering scathing criticism of the report's conclusion that the United States faces little immediate danger from acid rain, but much of their criticism was muted, since they feared that their research grants might be threatened if they were too outspoken. Just to illustrate the complexities of the debate, Eville Gorham of the University of Minnesota blamed the authors of the report for 'downplaying' the destruction of freshwater life by acid rain through a wrong definition of that term. He said that the threshold should be a pH of 5.5, not the pH of 5.0 that had actually been used as the criterion for damage to aquatic life. With this definition, 20 per cent of the lakes in the sensitive area, instead of the 10 per cent noted in the report, would have been classified as acidified. Michael Oppenheimer of the Environmental Defense

Fund points out that extensive research has demonstrated that freshwater organisms show serious reaction even at a pH level of 6.0. Some action has been taken: a Clean Air Bill contains an acid rain section calling for a 10 million ton reduction from 1980 levels in the emission of nitrogen oxides by 1997.[6]

Scientific journals have also criticized the report; witness an article with the title 'Report termed highly politicised'.[7] The writer says that the quality of NAPAP's work is generally considered to be very good indeed, and there is little to quarrel about so far as individual facts are concerned. But what is of concern is the way in which these facts have been used, perhaps out of context, to prove that the issue is not a serious one. It is this that has caused grave concern among the scientists. In neighbouring Canada, which probably bears the brunt of the burden from sulphurized emissions in the United States and the resultant acid rain, the environment minister, Tom McMillan, called the report 'voodoo science' designed to prove that 'the situation is not as bad as it is said to be and therefore there is no need to act with urgency'.[8] David Schindler of Environment Canada (the counterpart of the US EPA) is quite candid in his assessment of the report: '...it looks like a deliberate effort to downplay the effects but it is not due to the scientists involved... .'

In view of this conflict of views is it any wonder that, faced with exaggeration by the environmentalists and downplaying by the government, the public becomes utterly confused and is completely unable to get at the truth.

Acid rain is playing havoc

Let us try to assess the realities of the situation. Coal, it seems, is the main culprit behind the 'acid rain' phenomenon, and the position is liable to worsen now that coal is 'king' again. With the advent of 'cheap oil', coal was relegated to second place as a source of energy, but the drastic increase in the price of oil that occurred in the 1970s, together with a loss of credibility in nuclear energy as a source of power, has resulted in a resurgence of demand for coal. Worldwide, a 30 per cent increase in coal-fired generating capacity is expected between now and the turn of the present century. China, which has the world's largest coal reserves and has recently overtaken the Soviet Union as the world's largest producer, plans to double coal consumption by the year 2000. With its increasing thirst for power to cater for its ambitious industrial development programme, China may well become the world's largest 'polluter' in terms of sulphur dioxide and carbon emission.

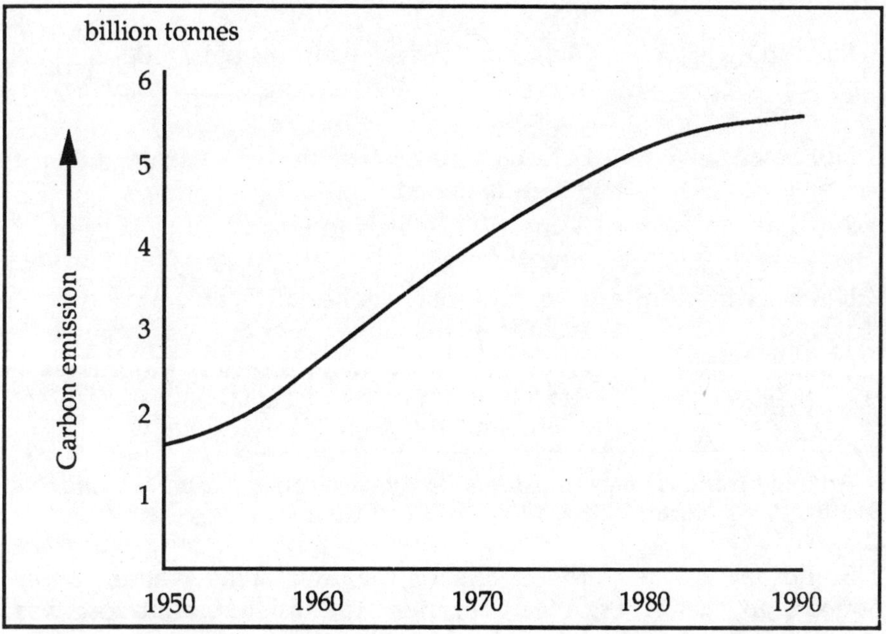

Figure 2.1 Carbon emission. Coal which releases more carbon than oil or natural gas is accelerating the rise in carbon emission.

Industrial activity, such as the use of coal and oil for energy, the use of petrol- and diesel-driven vehicles and the manufacture of steel are currently estimated to release into the atmosphere more than 5 billion tonnes of carbon and 0.1 billion tonnes of sulphur every year. Carbon emission tracks the world's energy needs very closely and the shift to coal in recent years, which releases more carbon than oil or natural gas, has further accelerated the rise in carbon emission, as illustrated in Figure 2.1. After a lull in the early 1980s, the rate of emission is now increasing at the rate of 100,000,000 tonnes per year. Emission is on the increase even in the United States with its strict pollution laws. The situation in the developing countries is even worse, where there are either no laws, or enforcement is extremely lax.

Table 2.1 Acid rain: extent of damage to lakes

Country	Extent of Damage
Canada	Ontario (1980). Some 140 lakes devoid of fish.
Finland	Helsinki (1984). Half of over 100 lakes surveyed highly acidified and may lose all fish.
Norway	Serious loss of fish in 175 lakes: another 900 threatened.
Sweden	Some 15,000 lakes too acidified to support aquatic life.
United Kingdom	Declining fish catches all over the country, including the Lake District.
United States (East)	Over 200 lakes in Adironack Mountains devoid of fish. Some 9,000 lakes threatened, one third of them seriously.

Adapted from *Worldwide Institute Report on Progress Toward a Sustainable Society*, W.W. Norton, USA, 1987.

Scandinavia was perhaps one of the first areas where it was noticed, in the 1960s, that acid rain and related pollution was causing the fish population to dwindle and that, in some lakes, aquatic life had completely disappeared. Sweden first reported lifeless lakes, soon followed by the other Scandinavian countries and the Northwestern USA, and also elsewhere in Europe (Table 2.1.). Later, damage to the forest areas was reported, the first serious serious warning coming from West Germany in 1982. It seems the damage is widespread, since in Europe alone some 20 countries have reported serious damage to their woodlands, more than 50 per cent being said to be affected in The Netherlands, Switzerland and West Germany (Table 2.2.). But the greatest threat must be to the developing countries, with their ever-increasing use of fossil fuels threatening not only the woodlands and the lakes, but the climate itself. No wonder, then, that apart from the vast volume of articles and papers on this theme in both the technical and the popular press, a number of full-length books have been published on this subject.

Typical of the titles that are appearing are *The Acid Rain Sourcebook* and *Acid Rain – Economic Assessment*.[9] It is also good to see a sound scientific approach to this emotive topic in the form of

Table 2.2 Extent of forest damage in Europe, assessed in 1986

Total forest area damaged %	Country
50–55	The Netherlands, Switzerland, United Kingdom, West Germany
35–40	Austria, Belgium, Czechoslovakia
25–30	Finland, France, Hungary, Luxembourg, Norway, Spain

Adapted from *Worldwide Institute Report on Progress Towards a Sustainable Society*, W.W. Norton, USA, 1987.

another full-fledged book by Wellburn.[10] The point he makes is that different biological mechanisms are affected by airborne pollution, each of the different pollutants gaining access to living organisms, including of course, human beings. He presents a global view of the subject and points out that, although the uptake mechanisms in plants and animals are very different, the underlying cellular events exhibit more similarities than differences.

The challenge of acid rain

An article in that prestigious American journal the *Scientific American*, with the thought-provoking title 'The challenge of acid rain', establishes without a doubt that acid rain is having the most disastrous effects and that there is an urgent need to rigorously control its causes.[11] It also describes some of the newer advances in technology which promise to bring environmentally and economically attractive solutions to this problem. Despite the growing evidence that acid rain constitutes large-scale interference in the vital biological and geochemical cycle of living things and their interaction with the environment, government action worldwide remains far too slow and lacks positive direction.

Thermal power stations are the largest single-industry user of coal and they have therefore been the subject of an enormous research and development effort in order to reduce the emission of the pollutants that cause acid rain. In the United States several technologies have been developed through joint industry-government collaboration, under the auspices of the Clean Act

Demonstration Program enacted in 1984. Three clean-coal technologies resulting from this programme are now being demonstrated in full-size plants:

Atmospheric fluidized bed combustion

With this system, pulverised coal and limestone are kept in a turbulent bed condition, allowing combustion to proceed at a lower and more even temperature than in the conventional boiler. This reduces the formation of the oxides of nitrogen, whilst the limestone captures the sulphur dioxide, thus attacking the problem at both fronts.

Pressurized fluidized bed combustion

This system is generally similar to that just described above, but the coal is burnt with compressed air. This increases plant efficiency.

Gasification: combined cycle

By reacting coal with steam and air at high temperatures, a gas consisting mainly of hydrogen and carbon monoxide is produced. This gas is burned and drives a turbine. The waste heat in the exhaust from the turbine serves to generate steam which then drives a steam turbine, thus generating further electricity. This cycle is extremely efficient and produces much less sulphur dioxide and nitrous oxides.

By incorporating such systems into existing plants that have no anti-pollution controls, the emission of sulphur dioxide can be halved. These new technologies are an attractive alternative not only on old and ageing plants which are doing so much to pollute the atmosphere, but also for new plants. This could be a quick and relatively cheap approach that would bring a substantial improvement, since most of the existing equipment would continue to be used. It is said that such a programme could ultimately reduce sulphur dioxide emission by more than 80 per cent, whilst the emission of the oxides of nitrogen could be reduced by some 50 per cent. Unfortunately the reduction, though ultimately dramatic, would be very slow, since such conversions of existing units will take a great deal of time.

The chemistry of acid rain

We have just laid much of the blame for acid rain and its consequences at the door of the power stations fired by the fossil fuels (coal and oil) and it seems that this is true. The term 'acid rain' was perhaps first coined by the British scientist Angus Smith to describe the precipitation in and near the industrial city of Manchester approximately 100 years ago. But now there are a multitude of 'Manchesters' scattered across the globe. With the rapid industrialization of first Europe, then the United States and Japan, acid rain has made its appearance in ever-increasing amounts everywhere. How does it happen? The gases given off when coal is burnt in boilers to raise steam includes SO_2 and NO_x (oxides of sulphur and nitrogen). The quantities emitted into the atmosphere up the chimney depends upon the sulphur content of the coal and the way in which the coal is burnt. These gases are then carried by the prevailing winds, combine with water in the atmosphere to make sulphuric and nitric acid, and then return to the earth, perhaps hundreds of miles away, in rain, snow, fog, or even in a dry state. These acids then enter the soil, affecting tree roots, or into the streams, rivers and lakes. Scientists are seeking to establish a direct relationship between the emissions and the effects, but whilst it is believed that there *is* a direct causal relationship, some scientists question this and wonder whether there are not other, hidden causes for the increased acidity.

The factors involved are many and very complex. Obviously the speed and direction of the prevailing winds, and their power of vertical transport, are important. Then there is the presence of other pollutants, the various chemical reactions that may take place and the efficiency with which the pollutants are finally removed from the air. There are some obvious contradictions which defy explanation. For instance, in the United States, although the emission of sulphur compounds has declined over the past few years, due to the increasing use of clean-up technologies, the ecological damage continues to increase. Perhaps we are dealing with a delayed response phenomenon, but no one really knows.

In any case, the problem remains. Should one seek to control the known pollution now, whatever it costs, or should one wait until the real causes of the damage and the mechanism is known? If one waits for that it might well be too late: the situation may reach a point of no return. In any event, who pays? The cost of the comprehensive clean-up facilities is phenomenal, as noted at the beginning of this chapter. Indeed, it could be that much of the money would be wasted – spent to prevent something which may not be doing the harm we thought it was. It is indeed this aspect of the matter which has caused a great deal of debate and friction. Whilst it seems an equitable principle that the polluter should pay, he gets no benefit: it is the community at large that benefits. The normally cordial relations between the United States and Canada have been seriously affected by the former's reluctance to launch an expensive clean-up programme in order to prevent continuing damage to Canadian lakes and forests. Similar problems have arisen in Europe, with the Scandinavian countries blaming West Germany and the UK for the increasing acidity of their waters and the damage to their forests.

The economics of acid rain

Environmental issues are becoming ever more important in the developed world and are now beginning to surface in the developing world as well. In the US presidential elections of 1988, such issues became a major point of debate between two of the contenders, George Bush and Michael Dukakis. The issues included clean air and water, acid rain and even waste disposal.[12] Both promised to enforce the laws governing toxic substances, ban the dumping of sludge in the ocean by 1991 and encourage the *recycling* of solid waste (the emphasis is ours) and launch a major campaign for waste reduction, minimizing spills and emissions.[13]

However, despite this political commitment, it is obvious enough that economics play a crucial role: cost is always a paramount factor. An international conference held in Washington DC in December 1984, sponsored by the Acid Rain Information Clearinghouse, was largely devoted to an economic assessment of the problems that were being met. The proceedings of this particular conference have been published as a book.[14] The conference brought together, on the same platform, economists, scientists and decision-makers, all seeking answers to such fundamental issues as:

- How can economics help to abate acid rain?
- Are there valid assumptions for an economic assessment of the issues involved?
- Is cost–benefit analysis valid for this highly complex problem?
- Is acid deposition linked to severe environmental degradation?
- How can this issue be legislated for, nationally and internationally?

Although numerous uncertainties remain, it is clear that cost is the single overriding factor inhibiting not only abatement but also legislation. The keynote address at the conference, which was given by I.M. Torrens of the OECD (Paris), highlighted the international perspective and thus posed a dilemma for the policy-makers. Accepting that enough is not really known to justify action and that it is most unlikely that further research over the next five years would provide all the answers, Torrens nevertheless pleaded for an action policy *now*. When it comes to the economics of the matter, where does the costing start and finish? What value should be placed on the physical and material damage that results? In one of the after-dinner speeches, Charles Elkins of the EPA, seeking to give an insight into the policy-making mechanics of the EPA, asked his audience: 'How would you quantify ... the value of the lakes which will be saved by the emissions reduction that you are costing out?' To answer this question, one has to know how many lakes will be saved by the control programme; then, we can ask the same question about the forests. Any estimates of this sort are patently mere guesses: how then can we ever arrive at the right decisions, if those decisions are to be governed by cost? In summing up, A.V. Kneese opined that estimates of the damage caused by acid rain, when known more accurately, could well be far greater than the present estimates. The only conclusion to be drawn, therefore, was that action was already overdue, and delay could only add to the ultimate price paid by the community.

It seems that there *are* viable technical solutions, and that the key problem is cost. Logically, the cost should be borne by the polluter, but the community at large will benefit and ultimately bear the cost as well. This means that there must be a strong political will and much mutual understanding and goodwill, if this problem is ever to be properly solved. For progress to be made, there *must* be a constructive partnership between the private sector and the government. Governments will have to provide fiscal and financial incentives, thus encouraging the development and use of superior technologies. A long-term perspective is required: there is no magic cure or 'quick fix'.

References

1 'Europeans do it to each other', *Economist*, **301**, 15 November 1986, p. 43.

INTRODUCTION

2 Mandelbaum, P., *Acid Rain – Economic Assessment*, Plenum Press, USA, 1985.
3 'Looking skyward: America's acid rain problem', *Chemocology* (lead article), Chemical Manufacturers' Association, Washington, DC, November 1986, pp. 2–4.
4 Ibid.
5 'A $94 billion boon to consumers?', *Wall Street Journal* (editorial), 23 September 1987, p. 34.
6 'Acid rain bill remains intact after House vote', *Chemical Marketing Report* (editorial), 7 March 1988, p. 23.
7 Roberts, L., 'Report termed highly politicised', *Science*, **237(4821)**, 18 September 1987, pp. 1404–6.
8 Shabecoff, P., 'Study criticized as "Voodoo" science', *New York Times*, 22 September 1987, pp. C1–2.
9 Elliott, T.G., Thomas, C. and Schweiger, R.G., *The Acid Rain Sourcebook*, McGraw-Hill, New York, 1984; Mandelbaum, P., *Acid Rain – Economic Assessment*, Plenum Press, USA, 1985.
10 Wellburn, A., *Air Pollution and Acid Rain – the Biological Impact*, Longman, London, 1988.
11 Mohnen, V.A., 'The challenge of acid rain', *Scientific American*, **259**, August 1988, pp. 14–22.
12 'The environment presidents', *Economist*, **309**, 8 October 1988, p. 36.
13 Chynoweth, E., 'The greening of the US industry', *American Chemical News*, 8–15 August 1988, pp. 18–20.
14 Shabecoff, op. cit.

Chapter 3

OUR SINK, THE SEA

Nearly three-quarters of our precious planet earth, is covered by oceans and, naturally enough, these vast reservoirs are seen as a logical place for the quiet disposal of the tremendous volumes of trash that results from our folly and misuse of valuable resources. But, if the waste materials dumped in the seas are disposed of foolishly, we may well be adding to our problems and simultaneously bequeathing a dreadful heritage to our successors. A banner headline in the *ISI Press Digest* proclaimed that we were 'awash in an ocean of trash', prefixing this headline with the declaration that it was a 'hot topic'.[1] We were presented with abstracts of three articles on this subject that had appeared in the popular press. The first was titled: 'The wavy waste'. The author said that the

> ...morning's tide swept a beautiful harvest from the sea over Port Bolivar's [Texas] beach: faded plastic bags, tangled ropes, rubber gloves, egg cartons, thousands of six-packs and soda cans, a plastic container of 'Milde Shampoo' from Euro Drogist in The Netherlands and a rusty refrigerator. Almost all of it was refuse dumped by ships at sea... .[2]

A beach may be delightful to look at, but what is happening under the surface of the water is another matter. Scuba divers talk of 'swimming through clouds of toilet paper and half-dissolved faeces' whilst fishermen bring in lobsters and crabs covered with mysterious 'burn holes' and fish whose fins are rotting away.[3] It was reported that, in the summer of 1987, some New Jersey beaches were closed for several days due to a 50-mile long garbage slick, caused perhaps by the illegal dumping of medical, construction and household refuse in the Atlantic Ocean by New York City.[4]

Apparently such dumping of waste is perfectly legal in the sense that there is no law against it – although there certainly ought to be. But how then could it possibly be enforced? It has been estimated that merchant ships around the world throw overboard some 100,000 tonnes of garbage per year and fishing vessels three times as much. Naval vessels are said to produce some two kilos of garbage per day for each crew member and, with some half a million crew members worldwide, that is a lot of garbage!

But ships are not the only source of refuse at sea. Bathers at Long Island, on the eastern American seaboard, were, naturally enough, alarmed to see needles, plastic tubing and phials washed ashore with the incoming tide.[5] This was obviously hospital waste and, with the continual threat of AIDS, the proper disposal of medical waste has become a most urgent issue. It seems that, so far as the United States is concerned, medical waste is not yet covered by the EPA's disposal regulations for hazardous waste – hence its appearance in the sea.

What is the effect of such indiscriminate disposal of garbage and trash? In a recent report by the US Congress's Office of Technology Assessment, it was reported: 'Hundreds of thousands of sea birds and an estimated 100,000 marine mammals die each year by ingesting or becoming entangled with plastic debris.'[6] It seems that the biggest culprits in such garbage is six-pack plastic holders and plastic rubbish sacks, which do not decompose. Such materials drift endlessly in the open ocean and plastic items some 50 years old have been found intact. Thousands of fur seals have been found wrapped and trapped in fish nets, whilst sea turtles often choke in an attempt to eat floating plastic sacks, under the impression that they are jelly fish. It seems that the world's oceans are seen as having an unlimited capacity to absorb waste, whilst the amount of waste committed to the oceans is considered to be relatively insignificant in such a context. However, the damage being caused by such waste is steadily becoming ever more obvious. It is an unfortunate fact that we rarely worry about a problem until it is almost too late. However, it appears that various 'safe options' are now being developed for the disposal of waste in the seas of the world.

Discriminative dumping

A number of companies have now set themselves up in the

business of accepting a wide range of waste materials, usually liquid, for disposal at sea. In the United States, dumping of waste material at sea is regulated by a range of different 'permits', as follows:[7]

Research permit

This is issued for a period of 18 months in respect of an otherwise prohibited material, allowing it to be dumped so that its impact on the marine environment can be studied.

General permit

This is issued to allow small amounts of non-toxic materials to be dumped, and is a one-off licence. There are no renewals.

Special permit

This is valid for three years and is renewable. It is designed to cover the disposal, on a regular basis, of material such as dredging spoil. It specifies the exact quantities and location for dumping.

Interim permit

This is issued for a maximum period of one year, is non-renewable, and is usually for industrial waste and sewage sludge or similar materials that exceed the permitted limits for normal disposal.

Emergency permit

This is another non-renewable permit, usually issued for highly toxic or radioactive materials which pose an unacceptable human health risk, whose disposal is not amenable to any other solution.

These prohibited materials include biological, chemical or radiological warfare agents; high-level radioactive waste; and plastic-like inert, synthetic or natural floating materials which could ultimately wash ashore. Materials prohibited in other than trace quantities include heavy metals such as mercury and cadmium; organo-halogens and other materials such as oils and greases, which if released in large quantities will float on the surface of the water.

The equipment and machinery required for the handling and disposal of industrial waste at sea is specialized and very expensive.

This has led to the establishment of contractors specializing in waste disposal at sea, with the appropriate expertise, which includes obtaining the necessary permits. By far the largest quantity of waste being dumped at sea is wastewater sludge, followed closely by industrial waste. Except for construction debris, the disposal of solid waste by ocean dumping is not as economical as landfill, nor would it normally be permitted. There is certainly an urgent need for the regulation of ocean dumping by International Convention, which would be binding on all the countries concerned.

Incineration: the safe option

The destruction of industrial waste by burning at sea was a logical improvement to dumping it there. This concept was pioneered by a German engineer, D. Sobinger, who was a specialist in combustion engineering. He started a company in 1967 which owned and operated a coastal tanker which had been converted to a sea-going incinerator. The highly dangerous chlorinated chemical waste products were burnt: chiefly the so-called 'EDC-tar', a black, liquid waste tar containing 60–70 per cent chlorine. On burning, this yields hydrochloric acid which, when condensed and entering the sea, was very highly diluted and was thus said to be innocuous. Following their initial success, the company purchased another ship, a freighter, which was also converted and put into service in 1972. Research work to measure the destruction efficiency of this particular disposal technique was first undertaken by Dr Kaeb of Bayer in 1971 and has continued ever since. The detailed nature of the research in this field is demonstrated in a book on the subject by Leo Spaans[8] who, since 1978, has been working for the North Sea Directorate of Rijkswaterstaat (The Netherlands), where he was engaged in Environment Assessment Studies. In this capacity he supervised, among other things, the research on board the incineration ships 'Vulcanus I' and 'Vulcanus II'. The Vulcanus I was brought into service in 1972 by Ocean Combustion Service BV, of Rotterdam (OCS). This ship was a true pioneer in the incineration of chlorinated wastes at sea and performed the first officially sanctioned ocean combustion of toxic wastes in the United States, successfully exceeding the environmental requirements of the EPA. This particular ship could handle 4,200 tonnes of chlorinated waste per year and proved so successful that the Shell Chemical Company were permitted to burn up to 50,000 tonnes of chlorinated organics

Figure 3.1 The Vulcanus II at sea. This vessel has the incinerators, seen at the rear of the ship. Only liquid waste can be burnt, and the chlorine content must not exceed some 70 weight %

and other waste by-products using this means. The Vulcanus II, which was specially designed and built for waste incineration at sea, came into service in 1982 and operates in the North Sea. This ship can be seen in Figure 3.1.

OCS is a Waste Management International Inc. company and one of their brochures, entitled 'Ocean Incineration', carries the subtitle: 'The ultimate way of disposing of liquid toxic chlorinated hydrocarbons and other organohalogen compounds'. It seems that there is no doubt that this group of companies are world leaders in the environmentally sound handling and destruction of hazardous waste. The key to their success, we feel, is the intensive, independent and steadily improving methods of research on the incineration process. It is now possible to obtain reliable data on the trace emissions from the incinerators and it has been proved beyond doubt that the destruction efficiency is normally more than 99.99 per cent, rather than the 99.9 per cent prescribed in the regulations. Analysis of the waste gas plume has shown that:

> Harmful effects of hydrochloric acid from the incineration at sea on life in and on the sea, on the beaches or on land can be ruled out... [the] biological effects of waste incineration on the North Sea ecosystems are below the detection level of the present day ecotoxicological field tests.

Although incineration at sea is seen as technically sound, it continues to attract political debate. In Spain the Environmental Ministry granted permission, despite opposition from the region's local government, for Vulcanus II to load 1,800 tonnes of toxic chemical waste for incineration at sea. Spain's Director General for the Environment has called the regional government's attitude 'incoherent', whereas the latter has called the central government's attitude 'infantile'. It seems that the sea is even more sacred to environmentalists than the land. Despite the fact that incineration at sea is extremely efficient and very effective, the European governments have nevertheless agreed that waste burning at sea should stop by 1994, as we noted earlier in Chapter 1. Unfortunately, the disposal options then left open may be worse, since burning this waste on land would be far more damaging than burning it at sea, and it still has to be disposed of.

The technique of waste incineration

Waste incineration is common enough, many public authorities using the process for the reduction of household waste, but waste incineration at sea becomes appropriate when toxic chemicals are involved. One OCS brochure with the title '15 years of waste incineration at sea' presents us with the history, the state of the art, its control and environmental impact, demonstrating that prolonged experience has resulted in the incineration of chemical wastes at sea emerging as a dependable method of destruction.The ships used in this service are under the full control of Germanischer Lloyd (FRG), Rijkswaterstaat (Netherlands) and the Coast Guard (USA), which are the relevant responsible organizations. When a ship leaves port for the purpose of incineration of waste it is under constant supervision until it returns, with regular warnings being given to any other ships that may be in the area. When the ship returns to port, all the monitoring data are collected on board by the relevant authorities.

It seems that the incineration of chemical wastes at sea is now a highly specialized and preferred technique for their safe destruction. Extensive research by independent research organizations and the inspecting authorities has confirmed that incineration at sea is not only economic but environmentally safe, doing no discernible damage to the land, lakes, rivers or marine life. The only major emission from incineration is hydrochloric acid, which is taken up partly by the alkaline seawater without any harmful effect, and is partly broken down in the atmosphere in an hour or so without any adverse effect such as, for example, acid rain. Incineration at sea has taken place in the North Sea, the Gulf of Mexico, the Pacific Ocean and Australian waters since the process was first brought into use in 1969, but by far the largest volume has been incinerated in the North Sea. The burning area is east of the Dogger Bank, far removed from any coasts. Since 1973 some 80–100,000 tonnes of waste have been incinerated annually. Not only is the process safe, but it is relatively cheap – far cheaper than incineration in a land-based incinerator, where hydrochloric acid gas would need to be neutralized with sodium hydroxide in expensive scrubbers. At sea this is unnecessary: the sea, being naturally alkaline, neutralizes the gas emissions at no cost whatever.

Incineration at sea in context

The incineration of some 90,000 tonnes per year of organo-chlorine wastes in the North Sea results in the emission of up to 50,000 tonnes of hydrochloric acid, most of which is taken up and neutralized by the sea. This input of hydrochloric acid does not cause any measurable chemical or biological effect. Yet from all the countries bordering the North Sea millions of tonnes of waste are pouring into the sea every year. In addition, as we saw in Chapter 2, many more millions of tonnes of sulphur dioxide and nitrogen oxides are released into the atmosphere above Western Europe every year, to end up as acid rain. Compared with all this, the amounts of hydrochloric acid emitted by incineration at sea are negligible and in any event, because they are emitted at sea, they are almost immediately rendered harmless. Many sources of pollution in the North Sea have been identified, but not yet quantified. What is very clear, however, is that the contribution of incineration at sea to pollution has been measured and has been seen to be negligible. Nonetheless, it is the continuing subject of controversy and debate. Whilst the technique is popular in Europe – toxic waste is steadily and safely being disposed of – in the United States the approach has been slower. The EPA has been studying the situation since 1974, with occasional test incinerations. Rules have been formulated and these have been the subject of public hearings.[9] However, a plan to launch a fleet of ocean-going incinerators to burn America's most hazardous liquid wastes was strongly opposed by the environmentalists, on the premise that the 'EPA is pushing a cheap – and hazardous – fix for a complex problem'.[10] It is feared that the quantities proposed for burning in the Gulf of Mexico are far greater than those currently being burnt in the North Sea, and hence its impact is thought to be unpredictable. An alternative site selected in the North Atlantic was found to be a breeding ground for whales, so that is hardly likely to be approved. The suggestion is that the EPA is not doing its 'homework', and is seeking to 'railroad' the proposal through, since it provides an effective solution to an ever-growing and intractable problem. The chemicals it is hoped to dispose of in this way include PCBs, cyanides, dioxins, herbicides and paint sludges.

Oceanic Society scientists fear that airborne emissions from incineration are having, and will have, an adverse and as yet unknown effect on the top layer of the ocean, which is seen as the initial key to most of the ocean's vast and complex food chain. A spokesman for another environmentalist group, Greenpeace, asks: 'Once it's out of sight offshore, who is to say what happens to the waste?' This is unfair, since there is no reason why the rigorous control procedures employed in relation to incineration in the North Sea could not be adopted in the United

States and elsewhere. High technology and data-logging under the immediate supervision of the appropriate authorities ensures that the waste is properly disposed of in accordance with the regulations. It is estimated that there will be some two million tonnes of liquid hazardous waste generated in the USA by 1990; incineration seems to be a safe means of disposal, and incineration at sea is much to be preferred to incineration on land.

Opposition to incineration at sea is now appearing in Europe as well. Some environmentalists consider the sea to be even more sacred than mother earth. According to an agreement signed in 1987 by the ministers of the countries bordering the North Sea, the only place in the world where toxic waste is currently being disposed of in this way (1988), this will have to cease by 1994. Whilst the pro-incineration lobby claims that only around 0.005 per cent of the toxic waste ever escapes, this statement is strongly challenged by Greenpeace, who point out that this ignores the emission of hydrochloric acid and its effects.[11] Research work at the Studsvik Institute in Sweden indicates that extremely minute quantities of waste could reach the plume gases emerging from the incinerator and react with the hydrochloric acid. Nothing is really known of the extent to which such chemicals could get into the plume, since it seems that it is too difficult and perhaps dangerous to measure them. Other researchers allege that they are, in any event, too diluted ever to be measured. Greenpeace's response to this is that they could well be there and will be absorbed into the microscopic film of organisms which live on the sea's surface, thus entering the food chain. The efficiency of the burn-up is also questioned, it being further alleged that the system of measurement ignores any new compounds that may be formed during combustion. Traces of such new compounds have indeed been detected in the furnace ash, but those supporting incineration retort that these traces are even smaller than those found in ordinary soils. The debate continues: who is to tell the truth of the matter, seeing that the concentrations in debate are too low to be measured by the instruments at present available? At the same time, concentrations much higher than can be measured today went unnoticed in earlier days. Is 'ignorance bliss'?

Unfortunately, most of the 'scientific' objections to incineration at sea that are being raised seem to be speculative, and in any event they will apply to incineration on land as well. What, then, is to be done? It is not really feasible to store such materials until a safer means of disposal has been developed. That ignores the risk inherent in storing such toxic chemicals. It is a maxim in relation to safety that 'what you haven't got can't hurt you', so surely the sooner such dangerous materials are disposed of the better.

References

1 'Awash in an ocean of trash', *ISI Press Digest* (feature article), 24 August 1987, p. 9.
2 Reinhold, R., 'The wavy waste', *New York Times*, 24 August 1987, pp. A1, A16.
3 Morgenthau, T., 'Beneath the surface', *Newsweek*, 1 August 1988, pp. 42–7.
4 Rangel, J., 'Federal brief describes illegal New York dumping', *New York Times*, 6 November 1988, p. 55.
5 Clark, M., 'Left with the tide', *Newsweek*, 20 July 1987, p. 56.
6 Berliant, A., 'Six-pack stranglers', *US News and World Report*, **103**, 6 July 1987, p. 72.
7 Conway, R.A. and Ross, R.D., *Handbook of Industrial Waste Disposal*, Van Nostrand, New York, 1980.
8 Spaans, L., *Incineration of Chlorinated Waste at Sea: process and emissions*, Foundation Marien Eco Consult, Scheveningen, The Netherlands, 1987.
9 'Waste disposal option: ocean incineration', *Chemecology*, (feature article), Chemical Manufacturers' Association, Washington, DC, April 1985, p. 6.
10 Cuneo, A., 'The uproar over burning toxic waste offshore', *Business Week*, 15 September 1985, pp. 124E–124H.
11 'Waste disposal – stinks ahoy', *Economist*, **307**, 21 May 1988, p. 100.

Part II
WASTE DISPOSAL AND TREATMENT

Chapter 4

WASTE DUMPS

As we come to consider waste disposal and treatment, there is no doubt that what is called 'landfill' is the most popular and most prevalent means of waste disposal. The dumping of waste at suitable sites around the country is the usual means of disposal worldwide. It may well be the quickest and the cheapest way to get rid of waste, but it is not necessarily a sound solution to the problem of waste disposal. More often than not it simply postpones the problem by taking the waste 'out of sight' and leaving it for some future generation to deal with. Indeed, at times, it has led to major disasters. The story of the Love Canal, which we have related at length elsewhere, and have mentioned in Chapter 1, provides a dire warning, in that the chemical waste that caused the trouble was very carefully and properly disposed of at the time. Although buried in the 1940s, long before the latest regulations came into force, the EPA said of the means adopted by the chemical company concerned:[1]

> [They] would have had no trouble complying with RCRA regulations. They may have had a little extra paperwork, but they wouldn't have had to change the way they disposed of the wastes.

The trouble was that the waste was disturbed – against the letter and spirit of a lease agreement whereby the land had been made over to the local authority – by the laying of sewers and other earthworks, thus releasing the poisons. People living in the immediate area eventually had to be evacuated and the story made media headlines. There was even a book written about it all, with the title *Laying Waste – The Poisoning of America by Toxic Chemicals*. Michael Brown, the author, drew attention to the fact that what had happened at Love Canal could easily happen elsewhere. Indeed, it was happening elsewhere: there were many similar unsafe dumps scattered across the American continent. Another site that received a great deal of media attention at this time was Swartz Creek, a site that was awash with toxic chemicals. It is now being cleaned up, but problems remain. It is said that 120,000 tonnes of contaminated soil from Swartz Creek have been scooped up and reburied elsewhere. But then what? That may well present yet another problem 50 years on – wherever it now lies!

The magnitude of the problem

Obviously, the stories from the Love Canal and Swartz Creek could be repeated many times over, not only in relation to the United States, but in many other countries. In the United States alone there is over 300 million tonnes of hazardous waste to be disposed of annually, of which about two-thirds derives from the chemical and petroleum industries. About 70 per cent of the total waste arising is disposed of as follows:[2]

		Per cent
1	By biological, chemical or physical means (such as incineration);	50
2	By disposal, such as injection wells, impoundments, landfill, etc.;	15
3	By storing, in tanks, waste piles, spoil heaps, containers.	35
		100

We propose to look in some detail at the implications of storing waste on the surface, presumably for later disposal, and its actual disposal as covered in (1) above. But it is very evident from the above table that half of the total waste produced remains somewhere to present a problem to posterity. This is in effect 'storing up trouble', as discussed below. In some of these cases, the waste may well have been treated to reduce its volume or to render it less toxic, but it is still there. There is no technical difficulty in making all such waste completely harmless: the only problem is the enormous cost involved. In response to the continuing public outcry and ever stricter regulations there is now much less waste being hidden away than was the case even 10 years ago, but it seems that little is being done to attack the problem at its roots – adopting processes that eliminate, or at least minimize, the amount of waste which must eventually be disposed of.

A national 'hit list' of waste dumps in the United States indicates that these are spread all over the country, but with the largest concentrations in the highly industrialized areas. It seems that the number of hazardous sites continues to grow, despite the fact that the government has authorized federal grants totalling billions of dollars for a major clean-up operation. There may well be more than 4,000 sites nationwide, with waste at times classified as toxic, ignitable, corrosive or dangerously reactive. There seems to be no stopping this continuous flow of dangerous waste. Factories making the multitude of appliances used in the home accumulate 'paint sludge'; there are the residues of chrome and nickel from metal plating shops; there are spent raw materials from the manufacture of paints, carpets and detergents. Then there are toxic and flammable solvents from the dry-cleaning industry, mercury from exhausted watch and appliance batteries, butane gas residues in the discarded cigarette lighter and lye left lying in the used can of oven cleaner. Indeed, the list is endless: the items we have just mentioned are merely typical.

What is the total? No one has the least idea. It is speculated, as mentioned above, that the United States generates some 300 million tonnes of hazardous waste annually, but whether one has to multiply this by two, or four, or even ten, to arrive at the quantity to be disposed of worldwide, no one knows. Over the 30 years from 1950 to 1980, the United States alone may well have disposed of some 6 billion tonnes of waste by way of landfill. Much of this material constitutes a hazard of some sort or another. As time passes, toxic chemicals may well be leached out to enter the water system, posing an ever-present danger to public health. There is the danger of cancer, birth defects, miscarriages, nervous disorders, blood diseases: such things have happened and are likely to go on happening. Can any country afford to ignore such a hidden danger? Public participation in the debate may help. The siting tactics for waste dumps have been discussed in a book dealing with the regulatory and institutional issues involved, whilst presenting acceptable methodologies for the disposal of hazardous waste.[3] This book is well worth study by those involved in such decisions.

Storing up trouble

There is no doubt that thousands of waste dump sites, scattered worldwide, carry deadly waste. The chemicals they are likely to carry include acids, pesticides, cyanides and even the extremely dangerous PCBs (polychlorinated biphenyls). They therefore pose a continuing threat. Whilst they are 'out of sight' they most certainly should not be 'out of mind'. That, it seems, was the trouble with the Love Canal. The chemical company who first deposited the toxic waste was fully aware of the potential danger, and did their best to ensure that the waste, once buried, was never disturbed. But unfortunately, the subsequent owners *did* disturb the waste. They either forgot (or perhaps ignored) the warnings that had been given. The trouble is that once a landfill dump is filled and closed, it is easily forgotten, and no thought is given to it until some trouble surfaces. By then, of course, it is often too late: the damage has already been done and the situation is often irretrievable. Of the many thousands of such sites worldwide, let us take a close look at just one abandoned dump, near Swartz Creek, Michigan, mentioned earlier in this chapter. We choose it primarily because the story is well documented.[4]

A creek that became a nightmare

In 1972 a certain Charles Berlin, together with a business partner, set up a hazardous waste incinerator adjacent to Swartz Creek, Michigan. As the business grew, and more and more chemical waste was entrusted to his company for disposal by burning, the incinerator was often overloaded, resulting in acrid dark smoke spreading across the countryside. What is more, since the burning was incomplete, there was corrosive fall-out from the smoke plume, playing havoc with cars' bodywork and reddening children's faces with a rash so severe that their eyes were often closed up. It is relevant to note that this particular waste disposal unit was set up when public awareness of the enviromental pollution problem was at its height: the seriousness of the situation was continually being emphasized through the media. Smoggy skies and murky streams were making headlines in the national press at the time.

The residents, confronted with this drastic fouling of their environment, and inspired by Verna Courtemanche, a retired mathematics teacher in her sixties, deluged the local state officials by letter and telephone, at the same time organizing rallies to arouse public consciousness. Yet it took nearly four years of intensive campaigning to get the dump and the incinerator closed permanently. By that time, of course, extensive damage had already been done to the local fauna and flora, and many of the residents had suffered greatly. But of course the toxic waste was still there, and it took Verna and her friends another 10-year fight before the government was persuaded to order a clean-up at its own expense. In the interim, not only had the health of many suffered, but house values had plummeted. Faced with legal action, Berlin and his partner abandoned the site and were declared bankrupt in 1980. When the clean-up started, some 50 trucks per day were moving spoil to another landfill site in Ohio. In all some 100,000 tonnes of earth were scraped up by bulldozers in order to remove toxic metals, spent motor oil, chemicals, drug and dye byproducts and other hazardous industrial waste. This took some three years, only to reveal five large storage tanks and over 30,000 drums of waste. The drums were rusting and breaking open, and some of the drummed material could not be burnt because of the toxic fumes that would result. And the question remains: what else is buried deep on this site?

Where had all this waste come from? Berlin and his partner were entrusted with this waste by chemical plants, automobile factories, steel mills, refineries and railroads. They had a thriving and expanding business, but it was part of their contract not only to remove the waste, but to dispose of it properly. This they had failed to do. In a holding pond nearby, the clean-up crew found over 1 million gallons of oily muck containing the dreaded PCBs. These biphenyls had been used in the past to stabilize hydraulic fluid, coolant for transformers and similar liquids, and eventually these had found their way to Berlin's dump. PCBs are very persistent and, when they get into the water, accumulate in fish, poisoning them. Through the fish they enter the food chain, the consequences of which were demonstrated horrifically at Minamata in Japan.[5] Another pond was found to contain drums of hydrochloric acid together with barrels of cyanide: leakage would have caused a cloud of lethal gas – hydrocyanic acid. Verna expressed her feelings, and that of the community, thus:

> We're prisoners, we can't sell our homes, we're afraid to drink from our wells, and out-of-town friends shy away from visits. My sister-in-law won't take gifts of my raspberry jam anymore.... At times,

> I've almost felt that addressing the problem of hazardous waste just makes it worse. You scream and holler, government acts, and easy answers elude you. But your only choice to solve the problem is to give it more attention, more effect. We're learning to do that, but I wonder if we're learning fast enough.

The cost of a partial clean-up at Swartz Creek has been over US$6 million. Fortunately some of the 200 companies whose waste was dumped there have accepted moral responsibility and even pledged in all some US$14 million. But tainted soil will remain, and many contaminants have now leached into the ground water. It is estimated that it will take several years just to map the groundwater pollution and to eliminate it may well take decades.

The disposal of industrial waste

Having seen how 'not to do it' let us see how it should be done. It is not wise to dump all waste together in any one place: it is far better to divide into what can be called 'safe' waste and 'hazardous' waste. Safe waste is that which has little or no potential for producing either dangerous leachate or harmful fall-out when stored. Hazardous waste comprises the remainder – toxic chemicals and the like. Any waste dump should comply with certain preconditions if it is to store waste safely:

- It should meet all the applicable regulations.
- It should be possible to maintain an aesthetic site.
- The site should be easy of access, at minimum cost.
- It must be possible to minimize litter and dust.
- It must be possible to protect the ground water.
- Workers at the site, users and the local community must not be put at risk.

The regulations and the standards required for waste disposal are constantly changing everywhere and it is therefore important to keep abreast of current legislation. The physical characteristics of the site should be determined in detail and a topographical map of the proposed site developed, with drainage patterns, type of vegetation, location of roads and utilities. The hydrological and climatic aspects of the site also need to be studied, with wind velocities and directions throughout the year. Of course, the nature of the waste to be disposed of is crucial to the decisions that are taken. In particular, it is essential to know whether there will be any leachate or gas generation. There is also a need to be clear as to the ultimate use to which the land may be put once dumping is complete. This is at least one valuable lesson that has come from the disaster at Love Canal. If it is likely to be used later for residential or recreational purposes, then special precautions are very necessary.

Human nature being what it is, there is a very natural tendency not to face reality and to sweep unwanted problems 'under the carpet'. This is what tends to happen with hazardous waste, with the result that every now and again we are confronted with a major health disaster. Such disasters ensure that care is then taken, but unfortunately the lesson does not seem to last. However, it does seem that conditions with respect to the disposal of industrial waste have been improving. There is much more public awareness than there used to be, and this is reflected in the laws and regulations that are being enacted. Certain minimum standards are now being made obligatory in more countries but, as each country will enact those rules and regulations which it sees as pertinent to its own special situation, we can hardly review them all. Let us then take the situation that has developed in the United States, perhaps one of the most highly regulated countries in the world. Since in that country there exists an open and free society, public reaction and opinion have done much to mould the rules and regulations that now prevail.

The Resource Conservation and Recovery Act (RCRA) was enacted in 1976 and, for the first time, EPA were directed to track, handle and monitor waste material right from creation to ultimate disposal. It was also required to supervise the operation of the licensed facilities, which had to be inspected regularly. The EPA finally had to ensure the safe closure of the site. As a result, the costs of the safe disposal of waste materials has risen considerably, and companies have been forced to review their operations in detail, striving to minimize the amount of waste being produced. They have also had to explore the possibilities for recycling, which is all to the good, as we shall see.

These regulations are designed to tackle the problem of waste at its roots – that is, prevention rather than cure. So far as the United States is concerned, the aspect of 'cure' has been introduced by the creation of a Superfund, which provides the money to clean up hazardous sites such as Swartz Creek and the hundreds of similar sites scattered across the country. Much of the activity in this direction is prompted and accelerated by activist groups of concerned citizens, seeking to preserve and protect their environment. We have mentioned one such, that grand old lady Verna Courtemanche, who led a campaign against the horrors of Swartz Creek. Despite the activities of such people, and the numerous laws, a vast amount of hazardous waste still continues to be disposed of carelessly and in unregulated dumps. Whilst ignorance may have something to do with this, it is certain that the primary factor is, and always will be, the cost of proper disposal. Currently, in the United States, some five million tonnes of industrial waste is discharged into, and carried away with, the domestic sewage per year. This includes most of the metal finishing industry's toxic waste, which is at present exempted by the EPA from more specific control. All this waste passes through city treatment plants, to be disposed of as sewage sludge, which is spread on the soil or dumped in the sea.

Not in my backyard

This famous phrase (NIMBY) sums up the attitude of most communities when waste dumps, waste disposal plants, waste treatment plants and the like are proposed. This is so, despite the fact that disposal and treatment plants, in particular, should really be welcomed since they offer a viable solution to a serious problem shared by us all. It is the community as a whole that is both directly and indirectly responsible for the waste, and proper treatment and disposal can only save us future harm and worry. Whilst it is true that some chemicals are deadly, even in minute quantities, the public at large seems to have developed what we might call 'chemophobia'. Anything that involves the handling of chemicals is feared and dreaded – quite unreasonably, since all such operations can be safely controlled and are carefully monitored. With the increasingly refined analysis techniques that are now available, extremely minute traces of chemicals in the soil, the water and the air can be recorded, and the alarm sounded should that be necessary. So the NIMBY attitude is largely unjustified and is also unfair. All these very necessary operations must be carried out somewhere.

The term 'NIMBY' has attained enough notoriety to warrant a technical paper carrying the word in its title.[6] This particular paper relates to a hazardous waste facility siting in Arizona State (USA): a state which is not heavily industrialized. The siting effort was based on the primary assumption that public opposition could be a major impediment to the siting of the facility: this is of course the spirit behind NIMBY. The basic principles behind the Arizona model were that state intervention in the form of legislative commitment was essential and that a public–private partnership is essential for the successful design, financing, construction, operation and maintenance of the facility.

Love Canal, Swartz Creek and their like are 'wounds' that have left 'scars' – indeed some seem to be wounds that will never heal. Consequently they are seen as symbolic of the perils of the industrial revolution and its accompanying pollution. Unfortunately, their clean-up proceeds at an agonisingly slow pace at an apparently astronomical cost.[7] For instance, at Love Canal, although the entire site has been covered with earth and plastic sheeting for a number of years now, there is dioxin-tainted mud from nearby creeks and some 2,000 barrels of waste still to be disposed of. So the problem remains, and it seems that it is difficult to find a satisfactory solution. It was proposed to build a new dump and related storage facility to safely contain the waste until it could be incinerated, but those who live nearby are afraid. Many of them are evacuees from the Love Canal area, and they think that incineration may be years away, so that they will remain at risk. They even fear the process of incineration, thinking that may well bring new hazards. Yet again, if an ostensibly safe dumping area is built, cost considerations could well lead to the site being used not only for the toxic waste from the Love Canal, but from other sites as well. Thus we find one of the residents, Louise Lewis, who remained in the neighbourhood through the Love Canal crisis, saying despairingly: 'They spent millions and millions of dollars to get rid of the chemicals ... now they want to put them back!' A Democrat representative for the Love Canal area, John J. LaFalce, adds that the EPA and other agencies involved have failed to learn the lessons from this particular tragedy. The plans being made by the EPA to handle the situation seem unlikely to encourage the former residents to return to the neighbourhood, which has degenerated into a slum.

We have been looking more particularly at the situation in the United States, where it seems that of perhaps 25,000 such waste dumps only a dozen or so have been cleaned up over some 10 years. EPA officials ascribe their slowness to the democratic public hearing process which must be gone through before concrete physical action can be taken. The position is further complicated by an unfortunate lack of trust by the general public in the EPA and other government agencies involved in waste disposal. But whilst we have taken some specific examples from the situation as it prevails in the United States to illustrate the problem, that problem is by no means confined to that country. It is a worldwide problem, and of course is most intense in industrialized countries. The handling of toxic waste is but a part of the landfill problem, which we review when we come to look at waste management. There we take up other aspects of this subject, such as 'garbology', the latest in-word in this context. What is very clear is that it is poor housekeeping, in particular by industry, that has accentuated an already difficult problem. As we come to consider the treatment and handling of waste and the relevant regulations concerning pollution and its control in the following chapters, we trust that we shall see the proper way to go.

References

1 Kharbanda, O.P. and Stallworthy, E.A., *Safety in the Chemical Industry*, Heinemann, UK, 1988.
2 Boraiko, A.A. and Ward, F., 'Storing up trouble – hazardous waste', *National Geographic Magazine* (cover story), **16**, March 1985, pp. 308–51.
3 Harthill, M., *Hazardous Waste Management – In Whose Backyard?, Westview Press, USA, 1984.*
4 Boraiko and Ward, op. cit.
5 Magnuson, E., 'A problem that cannot be buried', *Time*, 14 October 1985, pp. 24+ (5 pp.).
6 Weiss, N.L., 'Dealing with NIMBY', *Chemtech*, **18**, September 1988, pp. 540–4.
7 Smart, T., 'Love Canal – a new cleanup stirs old fears', *Business Week*, 31 August 1987, p. 30.

Chapter 5

THE LANDFILL PROBLEM

This chapter could well have been titled 'The landfill solution', since landfill is by far the most common and widely used method used for the disposal of all kinds of waste. Unfortunately, what was once seen as a satisfactory solution has now turned into a problem. In Chapter 4 we looked at the problems associated with waste dumps. The distinction between a dump and landfill is not really clear-cut. Landfill is an operation where the waste is used to fill up excavations or natural hollows in the ground. A dump may well serve this purpose, but not always. Waste can be dumped *on* the ground, as well as tipped into pits and covered, making it a sort of crude landfill.

The most common method

Landfill was, and still remains, by far the most common method used for waste disposal. Currently, landfill must be the largest repository of both municipal and industrial waste worldwide. Landfill and dumps are one of the oldest ways of disposing of waste and, despite their disadvantages, they remain by far the most common method in the United States and some other highly developed countries. Nearly 90 per cent of the refuse and solid waste is still buried in the United States, but by 1990 many of the landfills there may be filling up and more than half of the cities of America will have exhausted their landfills. Further, the dreaded contamination of ground water has been detected in some cases, intensifying the need to look for better alternatives. We shall look at the ground water problem in detail later. Incineration, discussed in the following chapter, is becoming more popular and the demand for waste-to-energy plants has been growing fast.

In the UK, too, perhaps 90 per cent of all waste, including toxic and other dangerous waste, is still disposed of by landfill. The most common method is the refilling of a spent quarry or gravel pit, the spoil being finally covered with a suitable soil, after which the area can be used for agriculture or leisure purposes. In some other countries, it still seems to be common practice to dump the waste on the ground, so that it eventually forms a hill.[1]

Landfill is undoubtedly a reasonably simple, cheap and effective way of disposing of large volumes of waste, although the general public tends to view it as a nuisance on account of the smells and the vermin, and the dust that can be raised, both during transport and at the site itself. This problem has been partially contained in recent years by the careful routing of the necessary traffic, and by using temporary surface coverings of soil or rubble as the filled area grows. The complex and sometimes toxic nature of the waste being handled can put the workers handling it at risk. Also, under certain conditions, anaerobic decomposition of the organic materials in the waste can lead to the emission of gases such as carbon dioxide and methane or even sometimes hydrogen sulphide and ammonia. Broadly speaking, this rarely presents any danger to the community, but it can have a nuisance value. Indeed, at some locations, the emission of methane has been put to good use: the gas is collected and used as fuel to generate electricity. The contamination of the ground water can also be a matter of concern, but research work, which we shall discuss below, has demonstrated that there are remarkedly few instances of this.

It seems that the bulk of the waste disposed of via landfill, whatever its origin (household, commercial or industrial), remains relatively inert after burial. However, most of the organic material in the waste biodegrades gradually over time, giving rise to a variety of products. The rate of decomposition is dictated by the nature of the waste, its pH, temperature and moisture content, the availability of oxygen and the bacteriological conditions. As a result of the various chemical reactions that occur as time passes, an aqueous layer, called leachate, is produced. This leachate is largely organic in nature, containing fatty acids, aldehydes and ketones, but it also contains some inorganic chemicals, such as the chlorides and sulphates of some of the common metals. This leachate is the most likely source of any contamination of the ground water. Landfill sites can be broadly classified by considering the behaviour of this leachate in relation to the geological factors obtaining at the site, thus:

1 sites where the waste and the leachate is fully contained;
2 sites where there is slow leachate migration, with significant reduction in its pollutant concentration;
3 sites where there is rapid leachate migration without any reduction in pollution.

The first situation is obviously by far the safest, and the most desirable. The means whereby the leachate is contained is usually natural as in, for instance, a disused clay pit. However, most landfill sites fall in the second category, and this requires some treatment of the waste. The suspended solids in the leachate are removed by filtration and the natural absorption and biodegradation processes will then bring about an appropriate reduction in pollutant concentrations. This approach, called the 'dilute and disperse' system, does pose some risk of damage to the ground water, but research, as discussed below, has demonstrated that there is no real cause for concern. Nevertheless, care needs to be taken even when the first solution, called the 'concentrate and contain' system, has been adopted, since if the waste becomes saturated with liquid, such as rainwater, the leachate may overflow, causing surface water pollution. This would then need some form of management: either the leachate would have to be removed for treatment elsewhere, or some form of leachate recycling system could be adopted, with redistribution over the existing contained area. The third type of site, whilst it can be used for innocuous waste, is obviously of no use at all for toxic waste.

The behaviour of waste in landfills

Some five years of intensive research work on 19 existing landfill sites by the UK Department of the Environment has provided some very useful guidelines in relation to waste disposal by landfill.[2] The prime objective of the research was hazardous waste and its behaviour, but at the same time a great deal of valuable information was gained with respect to the behaviour of landfill operations in general. The 19 sites were carefully selected from amongst the hundreds that could have been chosen, to ensure that they represented the main geological conditions encountered in the UK, and also that they were the repository for a wide variety of wastes. One of the sites studied included a contaminated area in the vicinity of an old gasworks.

The research involved drilling boreholes into and around the landfill sites and removing samples of waste material and the underlying strata, thus monitoring the condition of the waste and the quality of the ground water. Extensive analysis of the numerous samples taken yielded a broad picture of a three-dimensional spread of contaminants at each site. Physical analysis of the rock core gave an indication of the major hydrogeological controls that were being exercised over the movement of these contaminants. Water samples were drawn at regular intervals, in order to assess the variations that occurred in ground water quality over time. As a result of this detailed research we are now wiser as to the chemical and microbiological processes that occur in the waste, and about the relevant factors affecting the movement of leachate from the landfill into the ground water. We believe that the major conclusions resulting from this research are well worth summarizing here.

- Disposal of waste by landfill, under controlled conditions, is satisfactory and the ultra-cautious approach sometimes adopted is unnecessary.
- The presence of an unsaturated zone, about two metres thick, with pores and tissues that provide space for movement in the gas phase, at the base of the landfill, has been found to be extremely beneficial.
- An unsaturated zone (such as clay) slows down the rate of transport of the leachate and accelerates the biodegradation of organic materials, with the absorption and precipitation of inorganic materials.
- Where there is excessive liquid loading no unsaturated zone exists, and leachate is able to pass quickly to the strata below the landfill. The present knowledge on this subject (but see below) is not sufficient to pronounce on the suitability of a landfill site for a particular type of waste, but some guidelines can be given.
- Under normal conditions, with the occurrence of anaerobic degradation, certain metals, such as mercury, are converted to their relatively insoluble sulphides and are thereby retained in the waste.
- The disposal of domestic waste together with industrial waste seems to reduce the risk of ground water contamination. Even in the presence of a large proportion of industrial

waste, the leachate seemed to retain the characteristics of that arising from domestic waste alone.

The landfill sites studied had been, or were being, used for a wide variety of domestic and industrial waste, and particular attention was paid to what happened with industrial waste which contained acids, cyanides, heavy metals, oils, polychlorinated phenyls and other halogenated hydrocarbons. It seems that the presence of domestic refuse was always beneficial. Except when a significant amount of acid was present, when the metals were dissolved in the acid and passed into the leachate, it helped to retain heavy metals. Building rubble was found to have a greater neutralization capacity than domestic refuse, but had a smaller capacity for absorbing liquids. Domestic waste had the further advantage that it absorbed and retained oil, PCBs and solvents, whilst much of the cyanide disappeared harmlessly.

Because of its obvious importance, increasing attention is being focused on the composition and migration of ground water at such sites. A guide to the various monitoring techniques, and what a sampling programme should involve, is available.[3] Further, a regular annual symposium deals specifically with the engineering aspects of waste management and helps to integrate and liaise between the various disciplines concerned with the broad spectrum of waste management problems. The latest of these, the ninth in the series, and also the one a year earlier, focused attention on the geotechnical and geohydrological aspects of waste disposal. These are engineering areas of prime importance in the design and operation of waste disposal facilities, and various papers relating to these subjects have been brought together in a single volume by Lewis Publishers in the United States.[4] Earlier symposia in this series dealt with uranium mill tailings management, hazardous waste management and low-level waste management – subjects not immediately relevant here.

Landfills that have become notorious

It seems that it is in the United States that hazardous sites are gaining notoriety, but this may well be because of the powerful mass media there, which undertakes more investigative reporting than is usual in other countries. The list of hazardous sites in the United States is growing steadily and some of them have received a great deal of media attention. We have discussed two of them, those at Love Canal and Swartz Creek, earlier and these two, plus seven others, were featured in a cover story by the *National Geographic Magazine*.[5] Table 5.1 lists these nine sites, and it will be seen that they are spread right across the country. Whilst these particular sites are much talked about, providing ammunition for the environmentalists, there are plenty of landfill operations that are perfectly safe and can be used as a 'model' of the way in which it should be done.

Table 5.1 Priority hazardous waste sites in the United States

Site Location
Bridgeport (Pennsylvania)
Chem-dyne (Ohio)
Love Canal (New York State)
Price's Pit (New Jersey)
Rocky Mountain Arsenal (Colorado)
Seymour (Indiana)
Stringfellow Acid Pits (California)
Swartz Creek (Michigan)
Times Beach (Minnesota)

Based on data, given by A.A. Boraiko and F. Ward, 'Storing up trouble – hazardous waste', *National Geographic Magazine*, **16**, March 1985

Typical of a safe landfill operation is the use of a crater near Emelle, Alabama. This vast hole is in a layer of almost impregnable chalk some 200 metres thick. Hailed as the 'latest word' in landfills, it is proposed to dig some 20 pits which will eventually contain hundreds of thousands of tonnes of waste. Whilst this particular site is said to promise the safe containment of highly toxic waste for more than 10,000 years, it is still possible that the problem is only being passed on to posterity. However, disposal in this way is obviously far better than the means that some adopt. The irresponsible approach of some is well demonstrated by an example given by Dr Alan Block, a criminologist and research director of the New York State Select Committee on Crime. He tells of the 'typical' dodge of spraying to get rid of toxic waste via landfill: '[They spray] toxic waste on ordinary trash in conspiracy with bribed garbagemen ... compact the trash, send it to a clay landfill, and who'll know?' But to return to the proper way to do it. Another type of 'storage that works' consists of using the tunnels created by salt mining. These are used by the Kali und Salz AG of West Germany to store solid waste in drums at a depth of more than 60 metres. This facility is said to be deep enough, dry, and geologically stable, and now stores over 2 million drums. The storage system is so well organized that the exact location of each drum is precisely known, so that any possible future trouble can be localized and the relevant drums removed for treatment and disposal elsewhere.

The primary objective of proper, careful landfill is to prevent ground water pollution. Unfortunately, poorly handled landfill is by no means the only source of ground water pollution nowadays. Ground water is also threatened by highway de-icing salts, pesticide spraying by farmers, the application of fertilizers, and seepage from oil and septic tanks. Whilst it is estimated that these other sources of pollution only affect perhaps 1 per cent of the ground water system, the contamination usually occurs in just those places that have a high population density. The same applies to seepage from landfill sites: these sites are more often than not adjacent to population centres, since it is there that the waste is generated. As a consequence, we read that Atlantic City, in the United States, had to move its water wells in order to avoid toxic chemicals seeping from Price's Pit, a landfill area about a mile away from the city. The tap water had turned pale yellow in colour and aluminium saucepans were turning black due to the contaminated water. The problem with all such contamination is that no one knows what the ultimate effects may be. The immediate effects are often not too serious, but there may still be real long-term health effects. In the words of Dr Vernon Houk, director of the Center for Environmental Health at the National Center for Disease Control in Atlanta:

> Our skills in detecting toxic chemicals exceeds our ability to medically interpret what we find...risk assessment at most dump sites is somewhat less precise than a five-year weather forecast.[6]

It is very necessary, however, to be cautious in all such assessments, and not be blind to the realities. It does seem that sound judgements are not necessarily being made in this context, if we are to listen to the findings of one eminent scientist, Bruce Ames, chairman of the Department of Biochemistry at the University of California. A most respected authority on carcinogens, he devised, in the 1960s, a very simple and inexpensive carcinogen test using bacteria, hailed as a major scientific development and now being used extensively worldwide. In view of the grave public concern on this issue during the 1970s, he undertook extensive tests and found that a great many natural substances, such as fruit juices, brown mustard, celery and the like, proved positive as carcinogens or mutagens. His conclusion: 'man-made carcinogens are no more harmful than a peanut butter sandwich'.[7]

Correct landfill procedures

If the proper procedures are used, it is not only possible to dispose of wastes quite safely by landfill, but the environment can be protected as well. What should be done is well illustrated by an example in relation to the disposal of solid waste from a power station provided by W.H. Jansen.[8] He outlines the procedures adopted in relation to the waste from two coal-fired boilers on a power plant owned and operated by the East Kentucky Power Cooperative, and located near Maysville, Kentucky. The station consists of two units, a 300MW unit commissioned in August 1977 and a larger 500MW unit which started operations in March 1981. Only the second unit has an FGD (flue gas desulphurization) system, since the relevant regulations came into force well after the first unit was started up.

When the two units are operating at peak load, they discharge about 45 tonnes of dry ash per hour. In addition, the FGD scrubber on the second unit discharges some 50 tonnes of scrubber effluent per hour, containing 50 per cent solids. This waste is going to be produced at about this rate for some 35 years, the expected life of the plant. In all, that is estimated to amount to 350,000 tons of fly ash and other effluents on a 100 per cent solids basis. A most responsible approach was adopted to the disposal of all this waste. A suitable landfill area was identified nearby, where clay overlay rock at shallow depths. The entire strata was basically free of major folding or faulting and the leachate is unlikely to find its way into the ground water. Suitable parameters were established both for the landfill area and the necessary reservoirs and ponds.

An enviromental impact study confirmed that there would be no problem in the disposal of the solids waste from the power plant throughout its useful life. The entire landfill disposal system, including site development and the treatment plant, cost approximately US$30 million. Such a comprehensive and careful approach to the problem is to be commended, and this example has proved to be a model for other power plants in respect of the design, management and operation of pollution control facilities. Disposal of the dry fly ash could present a dust problem, but the FGD effluent is passed through a clarifier/thickener in order to increase the solids content from about 30 per cent to 50 per cent. The dry fly ash is then mixed with this effluent in a pug mill and the mixture transported to the nearby landfill area using bulldozers, scrapers and other earth-moving equipment. If required, lime can be added at this stage to create a pozzolanic reaction for stabilization. To avoid any dust problem at a later date, as the site dries out, surface covering is advisable prior to completion of the final soil cover. Run-off from the landfill area is trapped and retained, being returned to the power station for re-use. The topsoil is carefully stripped as the fill proceeds, stockpiled and then later used as cover soil. Seeding finishes the process of rehabilitation.

To ensure protection of the ground water, a water quality monitoring programme was put in hand well before work on the landfill area commenced. This is very important, since it provides a standard for later comparison. Water samples are taken at regular intervals, and this will continue right throughout the life of the project. As and when any leachate is detected in the ground water, corrective action will be taken as appropriate to minimize its production and prevent further contact with the ground water system. The utmost care needs to be taken in placing and compacting the clay blanket liner. The fill material has to be properly compacted and the final cover correctly placed and maintained. There are two types of run-offs, both of which need control: the one intercepted before it contacts the landfill area, and the other, from the landfill area itself. This latter may well be contaminated through its contact with the landfill material. What one needs to remember is that the waste from a power plant of this sort is relatively safe and non-toxic: yet an enormous amount of time and attention and substantial costs are involved if their safe disposal is to be ensured. With highly toxic wastes, of course, the problem is even more acute.

Handling toxic waste by landfill

Broadly speaking, landfill is used for the disposal of two distinct types of industrial waste: the non-toxic type, as discussed above and which is generally similar to municipal waste, and toxic chemical waste. If the latter has a significant heating value, it is usually burnt and the waste heat recovered. However, unfortunately incineration is not always possible. Toxic waste is usually in the form of a sludge which may yield undesirable leachate unless it is confined in a properly designed and really safe chemical landfill site which can hold chemical wastes in a variety of physical states. The site must be so designed that there is neither the possibility of leachate reaching the ground water, nor any significant risk of fire or explosion. Chemical toxic waste *can* be stored in such a landfill site without any adverse effect on the environment, and in some cases the wastes may biologically decompose into products which are perfectly safe. Where this is likely, the waste should be bulked but, in general, toxic material should be landfilled, stored in secure drums. The usual approach is to provide a concrete vault with a chemical barrier that ensures permanent burial, with no possible leakage of leachate. The main considerations in the design of a suitable chemical landfill are as follows.[9]

- It should be located close to the waste generation source and be large enough to cater for the waste expected to be created for a number of years.
- There should be a proper chemical barrier, designed to protect the people working in the area and in the nearby plant.
- The landfill should be so broken up that it is safe from fires, presents no risk to adjacent property and there is no odour nuisance.
- There must be careful segregation of wastes that may interreact.
- The waste should be so blended with the earth that gas generation is minimized, but whatever gas forms must be able to escape uniformly over the whole area of the landfill.

It is very unfortunate that with many of the older chemical waste landfill sites the hydrological problems were carefully assessed, but little attention was paid to geological problems. In particular, a permanent and absolute seal was not always provided, resulting in leachate contaminating the ground water. A satisfactory seal can be provided in one of three ways:

1 The use of an impermeable clay lining. Many areas have excellent clay soil that can be used for this purpose and it offers the most economic solution.
2 The use of an asphaltic membrane, similar to that used on roads. The trench is lined with asphalt beforehand and this has proved to be quite satisfactory.

3 The use of a plastic liner, usually 30-mill. thick PVC sheeting, the seams between the sheets being sealed as they are laid. This provides a sound seal, but it is rather expensive. Care must also be taken not to damage the sheeting as the waste is placed.

Since the area is sealed and liquid cannot drain away, it is essential to provide a duct or pipe to carry the leachate to an impoundment for any necessary treatment. The design should also be such that surface water is properly drained and not allowed to come into contact with the chemical waste or the leachate. Indiscriminate dumping of chemical waste in a landfill site is to be avoided. Each chemical waste is unique and it is best to analyse it before making a final decision as to the mode of disposal. For convenience, many of the major chemical companies have devised a system of classifying chemical waste in respect of the hazards they pose. The factors considered include health, flammability, stability and reactivity.

However, the problem is not finally resolved once the chemical waste has been consigned to a suitably designed landfill, since it is essential to monitor the landfill regularly. This can be done by means of test wells located around the landfill, as was done with the power station landfill site described above. The number of sampling points depends upon the expected variability of the various parameters and the degree of accuracy required. The distribution of the sample points has to take into account a number of factors: geologic, hydrologic, as discussed earlier in this chapter, and chemical. With fully confined chemical landfill sites of the type just described above, the primary check will be for the leakage of leachate which may contaminate ground water. The maintenance of clear and precise records is of the utmost importance. The exact location and quantity of the different types of chemical waste should be clearly marked on a map of the site. Once the landfill is complete, it should be closed safely – a design feature that must be in-built. Then the surface ground can be nicely landscaped, but the fact that it is a landfill area must never be forgotten – that was the primary lesson of the Love Canal!

References

1 Walters, J.K. and Wint, A. (eds), *Industrial Effluent Treatment*, Allied Science Publishers, UK, 1981.

2 Policy Review Committee, *Cooperative Programme of Research on the Behaviour of Hazardous Waste in Landfill Sites*, Final Report, Department of the Environment, HMSO, London, 1978.
3 Russell, D.L., 'Understanding ground water monitoring', *Chemical Engineering*, **84**, 26 October 1897, pp. 101–5.
4 Van Zyl, D.J.A. *et al.*, *Geotechnical and Geohydrological Aspects of Waste Management*, Lewis Publishers, USA, 1986.
5 Boraiko, A.A. and Ward, F., 'Storing up trouble – Hazardous waste', *National Geographic Magazine* (cover story), **16**, March 1985, pp. 308–51.
6 Anderson, K., 'Living dangerously with toxic wastes', *Time*, 14 October 1985, pp. 29+ (5 pp.).
7 Cone, M., 'Cancer chemicals – Are we going too far?', *Reader's Digest*, September 1988, pp. 120–2.
8 Jansen, W.H., 'Dealing with solid waste', *Asian National Development*, November–December 1986, pp. 38+ (5 pp.).
9 Conway, R.A. and Ross, R.D., *Handbook of Industrial Waste Disposal*, Van Nostrand, New York, 1980.

Chapter 6

INCINERATION

The incineration of waste would appear to be a very convenient and safe means of disposal. It has all the air of finality. The combustible material in the waste is converted to gaseous oxides, leaving an incombustible residue which should be non-toxic and quite suitable for disposal via landfill. What is more, the original volume of the waste is usually much reduced. At the same time, there is usually an opportunity to recover waste heat and use it profitably. Incineration furnaces can handle waste in a wide variety of forms: solid, paste, sludge, slurry, liquid and even gas, provided the appropriate design is used. In all cases, the objective is the same: to ensure the safe disposal of the waste. Unfortunately, safe disposal depends largely upon the way in which the incineration plant is operated. Due to the ignorance of operators, such plants are at times fed with materials which will not burn under the conditions obtaining in the incinerator, causing the waste to pass into the ash unchanged. At other times, the temperature maintained at the incinerator hearth is not sufficiently high to ensure complete combustion, and poisonous chemicals can then pass out into the atmosphere via the flue gas. There seems to be a general misconception in relation to the technique for maintaining proper combustion. It is thought that a large excess of air is desirable, but this is not so. A slight excess of air should be maintained to ensure complete combustion, but beyond a certain point it becomes counter-productive. Excess air can lead to cooling, with increased amounts of carbon monoxide and hydrocarbons in the flue gas and this is highly undesirable.

To set the subject into context, incineration has been on the increase in the United States. Table 6.1 sets out the cost–benefit of this type of operation.

Table 6.1 Increase in proportion of waste incinerated and cost, 1986–87

	Increase in quantity %	Increase in cost %
Incineration:	36	5
Resource recovery:	25	16–97
Landfilling:	5	10–48
Deep well injection:	1	–

Adapted from: J.M. Winton and L.A. Riach, 'Hazardous Waste Management – putting solutions into place', *Chemical Week*, 24 August 1988.

It will be seen that incineration is steadily increasing in use, whilst its cost is becoming ever more competitive.

Some general considerations

The incineration process as such is simple enough, involving the combustion of carbon and hydrogen, the two elements that are present in practically all combustible organic material. The other essential element is, of course, oxygen or air, and if the temperature is high enough combustion can be complete. The end-products of combustion are carbon dioxide and water, both in themselves harmless. The three basic but essential requirements for proper combustion are:

1 the residence time of the waste material in contact with air;
2 the degree of mixing between the air and the waste material;
3 the temperature of incineration.

These three requirements are not only essential in themselves, but they are interdependent. Thus, if any one of them is reduced, then one of the other requirements, or both, will have to be increased in order to maintain satisfactory combustion efficiency. Whilst the chemistry of the combustion process is quite straightforward, the significance of these three basic requirements for efficient combustion is often overlooked.

Whilst the basic combustion process centres round the conversion of carbon and hydrogen in the waste products to carbon dioxide and water, such waste also contains a host of other materials, combustible or otherwise. There can be sulphur, halogen compounds, phosphorus and inorganic salts. Some of these can be troublesome, since with the conditions created in the incinerator they can result in toxic compounds such as the sulphur oxides, chlorine and fluorine. However, if there is an excess of hydrogen some of these may well be hydrolized to safer compounds. In the absence of sufficient hydrogen, hydrolysis can be achieved, for instance, by steam injection, the steam breaking down at high temperature to provide further hydrogen. Solid wastes are in general more difficult to burn than either liquid or gaseous wastes. Their incineration is facilitated if they are finely divided, this ensuring that a much larger surface area is presented to the air for burning. In such cases, therefore, the shredding or pulverizing of the solid waste prior to feeding to the incinerator is strongly recommended.

If, as is often the case, the solid waste to be burnt has a low calorific value, then further fuel must be provided to ensure proper combustion. This fuel is not usually burnt together with the waste, but in a secondary chamber, which serves to increase the temperature of the combustion products from the burning of the waste. This ensures complete combustion. If the fuel were to be burnt with the waste, much more would be required, together with a lot of excess air, so that the combustion efficiency would be low. The two-stage process eliminates smoke and odour, and ensures that the flue gas does not pollute the atmosphere.

Incinerator design

It seems that there really is a future for incineration.[2] Integrated systems are becoming available with more advanced technologies and sophisticated computer controlled systems. This results in a trend toward more environmentally benign solutions to this particular approach to waste disposal. There are a wide range of alternatives available when we come to consider the type of incinerator that may be appropriate for a particular situation. Much depends not only on the character of the waste to be dealt with, but on the volume coming forward. The various types of incinerator in current use can be broadly classified in accordance with their basic design features, as follows.[3]

Static grate

This is used to burn solids and is ideally suited for small quantities of industrial waste. The system has low capital and operating costs, and is manually operated. After-burners can be used to ensure complete combustion, if required.

Multi-hearth

This type of incinerator is used for filter cakes and sludges. It consists of a number of horizontal circular hearths, each mounted above the other around a central tube. There is a set of radial arms over each hearth which rotate at a slow speed. The waste is fed in at the top and moves downwards, whilst the hot gases from the burning waste move upward.

Rotary

These have become popular for the handling of relatively small quantities of waste which is coming forward in several different forms: solid, slurry or liquid. It is in effect a rotary kiln: a refractory-lined cylinder, slightly inclined to the horizontal, that rotates at slow speed. The waste moves slowly down the kiln, being lifted and tumbled as it burns. The kiln can handle solid waste, sludge and even liquid waste, if introduced at the appropriate position along the kiln. It seems that rotary Portland cement kilns and aggregate kilns can use hazardous waste as fuel, thus fully recovering the energy value. This is in sharp contrast to the conventional incinerator which burns directly for destruction. Such kilns, it seems, are making more money from waste disposal than from aggregate handling, this giving rise to the question: what business are they really in?

Fluid bed

This system is particularly suited to the incineration of organic waste material, since it is in effect a process plant. The fluid bed system itself has three separate operating zones. Starting from the top, there is a 'free board' area, on which the solids from fluidizing drop back on the bed, then comes the fluid bed itself and finally at the bottom a windbox through which the air passes to fluidise the solids bed. The process is ideally suited to burn materials which can be fed at a controlled rate, such as slurries and sludges.

Multi-phase

This is used for bulky solids that cannot be processed in a rotary incinerator. In effect, the unit consists of two or more incinerators in sequence. The material with highest calorific value and with good burning characteristics is fired first. Its products of combustion, which are at a high temperature, are then used to promote the combustion of further waste material. Segregation in this way works much more efficiently than if all the material were burnt together, as it would be in a rotary kiln.

The processing of waste during incineration follows a logical course: separation, preparation, feeding, incineration and finally flue gas treatment. These steps are largely self-explanatory. Non-combustible materials should be separated from the waste before it is processed, since these only reduce the efficiency of operation. They are heated up to no purpose, and can also cause problems when they melt and form a heavy slag. Flue gas treatment should ensure proper air pollution control. There are a number of companies offering incineration plant, usually of standard design, but each situation should be analysed carefully to ensure that the appropriate equipment is chosen in terms of both size and type. Such companies often include in their advertizing material technical articles outlining the range of processes available, their capabilities and benefits.[4]

Domestic waste: to burn or not to burn

It seems that the appropriate disposal of domestic waste (popularly called trash in the United States) is now a burning question! It is being produced all over the world in ever-increasing quantities, but nobody really wants it. The citizens who produce it do not want it either dumped or burnt in their own backyard. Most of this waste is currently disposed of by landfill, but landfill sites are filling up fast. In the United States, which seems to have a headstart on all the other countries of the world in this context, it is said that nearly half of the metropolitan cities on the east coast will have no further landfill capacity by 1990.[5] At the moment a mere 5 per cent of such waste is incinerated, but this disposal process seems to be even more unpopular with the public than landfill when it is located at their doorstep.

As a consequence of the steadily reducing areas available for landfill, the cost of disposal by this means has soared. The residents of New Jersey, for instance, are now paying nearly 10 times as much for the disposal of their rubbish than they were paying only a year or so ago. The cost of disposal in Minneapolis has risen sixfold in six years. Local residents, particularly in New Jersey and Pennsylvania, are being bribed to accept landfill sites near where they live. Some cities, in desperation, have even been sending their rubbish abroad, and achieving considerable notoriety thereby. For instance, a barge *Mobro* with its tug *Break of Dawn* spent nearly two months in 1987 trying to get rid of about 3,100 tonnes of rubbish from New York around the Caribbean, but had to return home with its cargo intact.[6]

In view of the mounting problems associated with the disposal of household rubbish by landfill, one would have thought that incineration was an attractive proposition. The waste is reduced to ash, the volume being only one-tenth of the original. Furthermore, heat can be recovered to provide steam or electricity as a by-product. It seems it can be a profitable business, since a number of companies have grown up that charge cities for accepting waste for disposal by incineration, and of course charge for the electricity they generate and sell to local utilities. It seems that about 70 such plants were in operation, 20 under construction, and a further 100 were planned as of mid-1988 in the United States, but this is a pitifully small number when one considers the size of that country. However, incineration is already becoming widely used in many Western European countries. Unfortunately, even household waste incineration can bring its problems, chiefly because toxic gases can escape in the flue gas. These can be removed by scrubbing, but that costs money: the cost of disposal increases by some 50 per cent. Sometimes the trouble is with the ash: deadly toxins in the form of heavy metals and dioxins have been found in incinerator ash. This occurred in Philadelphia, necessitating the formation of an investigative task force. In the interim, some 28,000 tonnes of toxic ash was put on barges for disposal across the Caribbean. One of these, the *Khian Sea*, returned to base after a voyage to 'nowhere' in mid-1988, having disposed of less than one-quarter of its cargo. The barge was finally to be offloaded at Haiti until the authorities there realized what was happening and sent it away. The vessel was last sighted in the Indian Ocean after being denied landing rights in Sri Lanka. The only answer to what is otherwise an insoluble problem is to produce less rubbish and recycle more – issues to which we shall come later.

However, the story in relation to incineration is not all gloom and doom. There is a most interesting story concerning the development of waste-burning cookers in Zimbabwe.[7] A recent energy survey showed that a typical five-member rural household in that country used about 100 kg of wood per week for heating and cooking, or about a tonne per person per year. To help conserve the rapidly diminishing forests, a Zimbabwan farmer, Ian Hodgson, has devised a portable hollow-core cooker which can burn any kind of dry waste instead of wood. The design is based on a crude device that has been used for decades by farmers in South Africa. The original device consisted of a bottle placed in the centre of an upright, cylindrical oilcan round which leaves and straw were packed. The bottle was then removed and the compact waste lit through holes in the bottom of the can. The waste was found to burn slowly and steadily in the presence of the air that was drawn up through the hollow core. The cooker marketed by Hodgson copied this basic design in steel, with a steel core. It has been on sale since 1986 through a Harare-based company, Hollow Core Marketing Limited and has proved very suitable for baking, frying, heating the home, drying clothes and heating water for home use. Dry waste from more than 60 sources has been used with satisfactory results: this includes sawdust, straw, waste paper, leaves and peanut shells. The largest model, 90 cm high and 45 cm diameter, takes some 18 kg of waste and will burn for 48 hours before it needs refuelling. The real problem, of course, is to bring the device into large-scale use.

Incineration is certainly a viable alternative to landfill and there is renewed interest in this technique in the United States, in the light of the phased ban on the land disposal of toxic wastes. Nevertheless a mere 3 per cent of municipal waste is incinerated in that country at the present time (1988). Incineration projects are also going ahead in Europe, with a number of new plants, and many existing plants planning expansion.[8] The major chemical companies are increasing their in-house incineration capacity in order to conform to the stricter disposal regulations now coming into force. But, despite the growing capacity, one serious question remains: can land-based incineration plants satisfactorily destroy chlorinated waste?

Our review of incineration would not be complete without a mention of the electrochemical destruction of organic waste, which may well provide a satisfactory answer to this particular disposal problem in due course. This particular approach is currently (1989) at the pilot plant development stage in the AEA laboratories at Dounreay. This is a radically new means for the safe destruction of organically based toxic or hazardous wastes. It is likely to be particularly valuable for intractable wastes or for those wastes where the present disposal techniques are expensive. The process operates by generating a strongly oxidizing form of silver which breaks down the organic material to carbon dioxide and water. This is achieved at low temperatures and at atmospheric pressure. At the heart of the process is an electrochemical cell, which is in two parts, separated by a membrane. The anode cell contains nitric acid and a small quantity of silver salts. On passing current through the cell, the silver is changed from its normal valency to a strongly oxidizing state, which destroys the organic molecules, reducing them to carbon dioxide, waste and inorganic residues. The process is still being refined and systems for tackling solid wastes are also being developed. The next stage, hopefully, will be for a plant of this type to go into commercial production.

Landfill versus incineration

Whilst landfill at the moment seems by far the most common method of disposing of both municipal and industrial waste, and many prefer it because they see it as a 'tried and true solution', incineration may eventually be not only the preferred, but the cheaper, method of waste disposal.[9] The reason for this is the cost trend, illustrated in Figure 6.1. The projections look very promising, and from its present share of some 5 per cent of municipal waste, incineration may well be used for up to 40 per cent within the next 10 years. This is a consequence both of more efficient waste heat recovery processes, which improve the economics of the operation, and the increasing difficulty in finding appropriate landfill sites, which of course makes them evermore costly. For instance, Dudley E. Mecum, the president of Combustion Engineering's Urban Systems and Services Group, is reported as saying: 'In a densely populated state it is difficult to site a landfill within cost-effective distance of the community it serves.'[9] Another factor which has served to encourage the use of incineration is the fast growth of the commercial incineration business. Companies in this field are now prepared not only to supply, but also to operate and maintain the plants. This is very attractive to municipalities, who lack the necessary expertise.

The two major types of incineration plant currently used are 'mass burning' units and refuse-derived fuel (RDF) units. With mass burning, everything goes into the furnace, including grass clippings, milk cartons and even refrigerators. With the RDF unit,

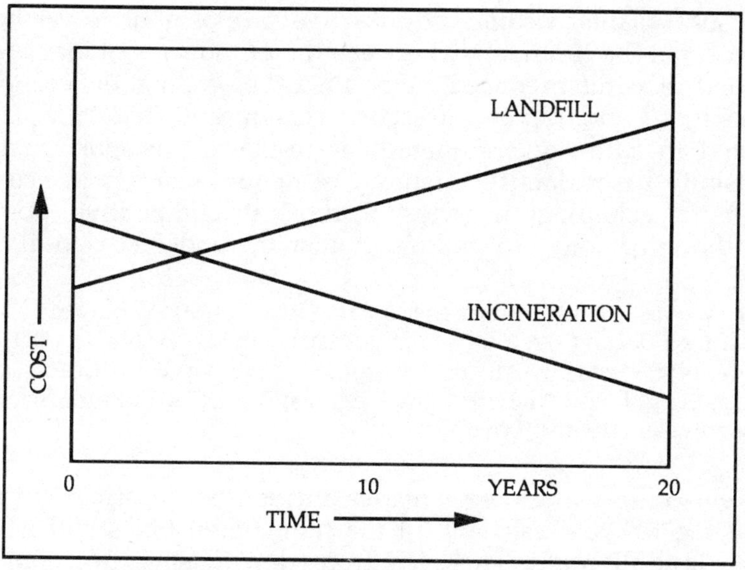

Figure 6.1 **The cost of waste disposal. This graph compares the possible cost of the transfer and disposal of waste via landfill and incineration over the next 20 years. The actual cost will vary from location to location and from country to country, so the graph merely indicates the trend. The cost in US$/tonne is not directly relevant, but the difference, which was about US$10 per tonne in favour of landfill, may increase to US$60 per tonne in favour of incineration.**

the non-combustibles are removed, with the result that the feed to the furnace is more uniform in quality. This means that the rate of production of steam can be more closely related to demand, since it responds more promptly to the firing rate. In both cases, a fairly high temperature is maintained in the incinerator so as to ensure the destruction of toxic gases, and the flue gases may well pass through a scrubbing system for additional protection. Of course, the residual ash still has to be disposed of by landfill, unless some use can be found for it.

The economics of incineration depend to some extent on the location of the plant, but are heavily dependent upon the price that can be secured for the steam or the electricity generated by the plant. Where the low-pressure steam normally generated by such plants can be used effectively for, say, domestic or factory heating, or in the canning of vegetables or concentrating orange juice, such plants become an attractive proposition. Nevertheless, it seems that the controversy between the proponents of incineration and landfill as the preferred method is never-ending: the economics are so loose and rather uncertain. This means that they can be adjusted to suit any argument. For instance, the spokesman for a Pittsburg-based landfill disposal company, which had acquired mass-burn technology in order to protect its competitive position, even though it did not believe in that method of disposal, said:

> The whole world thinks this incineration is nice, but the energy produced doesn't even pay for the salaries of the employees, let alone the capital costs, which are awesome ... We serve the wants of the people, right now they're biased against landfills, thinking each one is equivalent to the Love Canal.[10]

He went on to make the point that incineration has its own risks, whilst the residual ash and all the non-combustibles still have to go the landfill route. Those confronted with such decisions will have to make their own choice, but this should not be done until all the 'pros and cons' have been carefully assessed and studied in depth.

Pyrolysis: a novel approach

Whilst the process of pyrolysis reduces the waste by decomposition through heating, it is not a process of incineration. The waste is heated to a high temperature in an oxygen-deficient atmosphere in a retort: this results in both physical and chemical decomposition. The prime product is a solid which may be described as a 'coke', readily firable. This can be stored until used, and herein lies the chief advantage of this particular process over direct incineration, where the released energy has to be used forthwith. There are a number of texts dealing in some detail with the pyrolysis process.[11] Pyrolysis is described as a thermal treatment process and, whereas incineration is an end in itself, pyrolysis is only the means to an end. Pyrolysis results in what may be described as a safe to handle 'crude fuel' which can be used either at the same location where it is produced, or elsewhere. Unlike incineration, waste for pyrolysis requires fairly rigorous pre-sorting to remove non-combustible items such as metals and glass, and size reduction. The hydrocarbons in the waste are broken down into lower molecular weight hydrocarbons, the degree to which this is achieved depending upon the rate of heating and the temperature of firing. It seems that some of the constituents in the waste act as a catalyst in this process, but their precise role is not known, nor can it be predicted. The main benefits to come from using pyrolysis for waste reduction are as follows.[12]

- It is a resource recovery process without any major release of heat.
- The volume of gases produced is considerably reduced, thus bringing large savings in power and gas cleaning requirements, whilst pollution is minimized.
- The residue is innocuous, sterile, and in friable form.
- The products can be easily handled and transported.
- A convenient and refined fuel is produced which can be burnt in conventional boilers.
- A pyrolysis plant is more compact and cheaper in terms of capital cost than an incinerator capable of dealing with the same volume of waste.

It seems that the pyrolysis process is coming into favour and commercially viable plants have been built in many parts of the world. The process is particularly valuable in dealing with waste plastics and rubber, the scourge of modern civilization. Valuable economic and technical data on the process is being steadily accumulated and it is of growing importance in the field of waste destruction. The plants are reliable in operation, despite their sophisticated design, and the development of standardized modules has helped to reduce their cost.

To take a brief look at the pyrolysis process itself, the operating temperature is usually around 800°C, substantially lower than that maintained in the average incineration plant. The basic equipment consists of a refractory-lined combustion chamber, loaded with waste, which has a limited flow of air to it. The initial heating is by oil or gas burners, but the reaction is autothermic – that is, self-sustaining as to temperature once the appropriate operating temperature has been reached. The waste decomposes under quiescent conditions and there is very little carry-over of particulate matter. The flue gases, with such particulate matter as is carried over, pass through another chamber located immediately above the combustion chamber, where they are mixed with air and the temperature raised to about 1200°C. This ensures the complete burn-out of smoke. Experience with pyrolysis installations has shown that a volume reduction of some 90 per cent can be achieved, as with incineration, whilst the process can handle difficult wastes with the minimum of air pollution. There are several different pyrolysis processes on the market, with others in the course of development, and this process seems to have a promising future in the field of waste disposal.

References

1 Winton, J.M. and Riach, L.A., 'Hazardous waste management – putting solutions into place', *Chemical Week*, 24 August 1988, pp. 26+ (15pp.).

2 'Incineration – A burning issue', *Chemical Business*, (editorial) August 1988, pp. 12+ (2 pp.).

3 Walters, J.K. and Wint, A. (eds) *Industrial Effluent Treatment*, Allied Science Publishers, UK, 1981.

4 Cegielski, J.M. Jr., *Hazardous Waste Disposal by Thermal Oxidation*, John Zink Company, Tulsa, Oklahoma, 1981.

5 'Rubbish – burning question', *Economist*, **307**, 28 May 1988, pp. 39+ (3 pp.).

6 Anderson, H., 'The global poison trade – How toxic waste is dumped on the Third World', *Newsweek*, 7 November 1988, pp. 8–11.

7 'Burning waste, not wood', *Newsweek*, 14 March 1988, p. 3.

8 Hunter, D., 'The burning issue of waste in Europe', *Chemical Week*, 15 June 1988, pp. 19–20.

9 Brooks, K., 'Garbage disposal: to burn or not to burn', *Chemical Week*, 11 June 1986, pp. 44+ (2 pp.).

10 Ibid.

11 Conway, R.A. and Ross, R.D., *Handbook of Industrial Waste Disposal*, Van Nostrand, New York, 1980, Ch.10; Kut, D. and Hare, G., *Water Recycling and Energy Conservation*, John Wiley, USA, 1981; Winton and Riach, op. cit.

12 Winton and Riach, op. cit.

Chapter 7

GOING EVER DEEPER

To bury or to burn? This is a continuing controversy, and the answer to this burning question can only be determined case by case. There is no single or simple answer, although we believe that incineration must be the preferred alternative whenever that is possible. However, in some circumstances, burying the waste may still be the preferred alternative, if it can be buried deep enough. But how deep is deep enough?

Deep well injection

Deep well injection is a technique for storing waste permanently in underground strata. It is a viable method of disposal where circumstances permit, since it drastically reduces, if not eliminates, the possibility of contamination of surface water or ground water by the toxic waste or its leachate. Disposal in deep wells is a tried and proven technique in relation to oil production. It is quite a common practice, when large quantities of saline water come up with the oil, to return this water to the subsurface whence it came. This subject has received rather limited attention in relation to waste disposal, not only because of the technical problems, but because of the legal and regulatory restraints and uncertainties that exist. The uncertainties relate not so much to the ability to dispose of waste in this way, but to the long-term consequences. Nevertheless we feel that the subject is important enough, and that there has been sufficient work done on it, for us to review this particular approach to waste disposal in some detail. A basic step is to determine whether the proposed injection stratum is suitable for the purpose. Some of the essential requirements for a suitable injection stratum are as follows:[1]

- There should be sufficient capacity, sufficient depth and porosity (degree to which the voids are already occupied by connate water).
- Aquicludes should be present (impervious strata that isolate the injection area from underground water supplies and mineral resources).
- Permeability has to be sufficient to accept the waste at the required rate.
- The fluids already there have to be compatible with the waste being injected. Alternatively, the waste can be treated to ensure compatibility.
- A proper geological survey should be made, providing data on the injection zones and confining beds.
- An operational plan should be formulated for the injection procedure, with data on the pressure levels, the source and the characteristics of the fluids involved.
- Detailed engineering data should be available on the proposed well casing and its cementing.

An appropriate stratum for injection is usually limestone or sandstone, whilst the aquicludes can be clay, marl, limestone, siltstone, sandstone, or gypsum. Much preliminary testing is required to ensure satisfactory operation. There should be pump-in tests to determine the rate of flow at various injection pressures. It is also advisable to carry out tests for compatibility of the waste with the particular stratum, preferably at the temperature and pressure which prevails at the bottom of the hole. The importance of such testing is well illustrated by the story of a well that had to be abandoned after only three years in operation, with a loss of some US$2 million: the capital invested in the well.[2] It seems that there were unstable organic compounds in the waste that reacted with the very saline water in the aquifer. This led to the precipitation of iron carbonate and iron hydroxide, which plugged the injection wells.

In sharp contrast to this unhappy story, there are many success stories. For instance, chemical plant waste was injected into isolated salt water sands nearly a kilometre below potable aquifers for over five years without any problem. But great care was taken, the waste being pretreated. As part of the pretreatment, its pH was adjusted to between 5 and 8: then the suspended solids present were much reduced by sedimentation, skimming and filtration through deep-bed sand filters. A pH of 5 ensures that there will be no corrosion of the carbon steel used for the piping and other plant in the system, whilst the top limit of 8 ensures that there will be no undue swelling of the clays in the injection zone. This case illustrates the supreme importance attached to limiting the amount of suspended solids and controlling the quality of the waste so that there is no possibility of plugging the porous stratum into which the waste is being injected.

Despite its apparent effectiveness, there is much opposition to the disposal of waste by deep well injection. The legislation and the regulations relevant to this method of disposal are still being developed, and environmentalists are seeking to ensure that they are tightened up. A senior attorney with the US Natural Resources Defense Council has been reported as saying:

> We're certainly working to be sure that injection is treated on an equal footing [to landfill] as EPA decides whether it's a safe practise to continue the true out-of-sight, out-of-mind technology. We'd like to see it greatly restricted.[3]

From what we have already said about the disposal of waste using deep well injection, it is very clear that not all waste is suitable for disposal in this way. But, if the geological conditions are favourable and the strata appropriate, it seems that this method is a quite appropriate and safe way of disposing of certain types of waste, despite the reservations of the environmentalists.

An estimated 60 per cent of all industrial toxic waste in the United States is being disposed of by injection in deep wells. It is certainly big business: one company active in this field, Chemical Waste Management, has actually been described as the 'world's largest hazardous waste disposal firm'.[4] Typical of the injection wells operated by this particular company is a 1.5 km deep well near Corpus Christi, Texas. Huge quantities of acids, caustics and toxic solvents have been injected between layers of rock in an ancient seabed without any apparent ill-effects – so far!

It seems that therein lies the problem. Despite the apparent success of this procedure, the environmentalists are still suspicious and the demand for deep well injection seems to be declining. This may be due in part to the protests of environmentalist agencies, such as the NRDC (National Resource Development Corporation) in the United States. It is particularly concerned that, because of the EPA rule prohibiting the land disposal of selected wastes, a major loophole may be created in sanctioning migration of waste out of injection zones. The NRDC has filed a suit against the EPA, objecting to their stand on underground injection. It is particularly concerned about a new injection facility in Kern County, California, designed to dispose of 750 million gallons per year of hazardous waste – the largest such installation to date. The basis of their objection is that the seismic geology at the site is unsuitable.[5]

Deep well injection: advantages over incineration

Has deep well injection any advantages over incineration as a method for disposing of hazardous waste? T.H. Maugh, in an article which is the third of a series of four on the proper disposal of hazardous waste, puts it like this: 'Incineration destroys hazardous wastes, but is expensive: deep-well injection is much cheaper, but very controversial'.[6] Why? The problem seems to be that the safety of deep well injection cannot be clearly demonstrated. It *appears* to be safe enough, but there is no solid proof. Incineration has a poor image: it conjures up a vision of a municipal plant spewing out dark smoke with its accompanying pollutants. In actual fact, with the proper controls, incineration is one of the safest and cleanest methods of disposing of waste – especially hazardous waste. Unfortunately, and perhaps a little unexpectedly, it is at the moment also one of the most expensive. Current regulations require a minimum of 99.99 per cent destruction, which demands sophisticated technology and elaborate equipment: hence the high cost of disposal by this means.

On the other hand, deep well injection is a familiar technique, and the waste, once placed some 1–3 km below the earth's surface, is assumed to be safely stored there, in perpetuity. At least it is out of sight! The technique has been used for the disposal of brine from oilfield activity since the 1920s, and for the disposal of hazardous waste since the 1950s. It is fairly freely available, since in the United States for example, suitable sedimentary formations for deep well injection are said to occur under about half the total land area. Most of the wells used for waste disposal are captive: that is, they are owned by the company who produces and has to dispose of its waste. However, some are owned by companies in the business of waste disposal, particularly in the United States. The safety of the operation is said to be demonstrated by the fact that most of the injection sites already carry brine, which is clearly constantly separated from freshwater zones by the relevant aquicludes (described above): so why should not the same apply to the waste that is deposited there? With proper care during construction of the well to ensure that the strata are not disturbed, injection of wastes into such wells should be perfectly harmless. Further, the situation is not irreversible. If at some future date technical innovation should result in some use being found for the waste, it can be pumped back to the surface for re-use or recycling, subjects which we shall be discussing later.

So far as cost is concerned, the main cost element is the drilling of the well itself: operating costs are usually almost negligible. In some cases even pumping is not required; negative pressure at the well head can be used to suck the waste down into the depths. As with all the waste disposal techniques, deep well injection has had its critics over the years. Some of the objections raised against it are:

- lack of precise knowledge as to the fate of the injected waste;
- the possibility that injection in deep wells may have given rise to earthquakes;
- earthquakes are liable to release the waste.

We would venture to suggest that sufficient experience has been gained over the years to make deep well injection a very safe disposal technique, provided that the construction and processing of the waste is properly carried out in accordance with the regulations. Certainly its popularity has been growing and the share it takes of hazardous waste is likely to increase, since the cost of the other acceptable alternatives is likely to increase much faster than the cost of deep well injection.

Deep sea waste disposal

So far we have been considering what happens on land, but we should not forget the oceans. We have already seen, when considering 'our sink the sea' (Chapter 3) that the oceans offer enormous scope for the disposal of industrial waste. This aspect of the subject is of such importance and is so vast in scope and concept that it has merited six full-length volumes under the general title *Wastes in the Ocean* by panels of experts. The breadth of treatment accorded to this particular subject is well illustrated by the titles of the books in this particular series:

Vol 1 Industrial and sewage wastes in the ocean
Vol 2 Dredged material disposal in the ocean
Vol 3 Radioactive wastes and the ocean
Vol 4 Energy wastes in the ocean
Vol 5 Deep sea waste disposal
Vol 6 Nearshore waste disposal

This series represents a most valuable addition to the literature on this vast subject. They are part of the Wiley Interscience series of texts and monographs on Environmental Science and Technology. In our present context, it is Volume 5 that is of particular interest.[7] We are presented with an authoritative treatment of deep sea waste disposal, its effects on, and interaction with, the marine environment, together with the economics of the subject. The publisher describes this book as presenting 'the most current information available on a growing problem'. Seen once as merely a repository for sunken ships, the oceans are now receiving ever-increasing amounts of toxic materials: it is this that creates the problem.

It is certainly true that, as the problems associated with the disposal of waste on land become ever more acute, increasing attention is being paid to possible disposal at sea. At first sight, the wide, deep ocean appears to offer a safe disposal site, but is this really true? It is clear that some regions of the open sea present a better choice than others for the purpose of waste disposal because of the differences in water circulation, the chemical conditions, and the biological productivity of the waters. But what happens when waste is dumped on the deep ocean floor? It seems likely that such waste is highly diluted and dispersed in the ocean, and can be assimilated by natural processes without detriment to marine life. After all, it is said, all sorts of things have been dumped into the sea for thousands of years without any apparent ill-effect. Nevertheless we must recognize that, despite its size, the ocean is still a limited resource, so that much will depend upon the *volume* of material deposited. Research has already demonstrated that man *is* bringing about global changes, and this may well apply to the oceans, even though the impact of his actions is not immediately apparent.

One serious difficulty when appraising the dumping of waste in the deep ocean is to devise a method of verifying what is happening. Subtle but significant changes in the marine ecosystem can occur without their being recognized. It is true that the deep ocean has a much larger capacity in this respect than rivers or coastal waters, where the detrimental effects can be seen and corrective action taken, but that is not to say that similar results will not eventually be brought about in the deep ocean. It is naive, to think that, just because the ocean is vast in extent, and remote from human activity, we do not need to worry about what happens there. The ozone layer is remote, but man is beginning to be very concerned about what is happening there. Much is known about the ocean, but there is a great deal more still to be learnt.

Despite this ignorance, substantial amounts of industrial waste are being disposed of in the deep ocean. Much of it is waste acids from steel mills, refinery waste and pesticides. The usual manner of disposal is by barge to a permitted area. Drums or other containers carrying waste are weighted, to ensure that they sink to the bottom. All such disposal is rigidly controlled and must be certificated. The regulations govern not only the disposal site, but the manner of shipment and discharge. Barges have to be double-skinned, and the method of discharge at site is typically from 4–20 tonnes per minute through hoses that are trailed behind the barge at a depth of some five metres. It is obvious that the purpose of the regulations is not to protect the disposal site, but rather to ensure that disposal is effected at the designated site and nowhere else.

Most of the countries in the developed world have regulations governing the disposal of waste at sea. Numerous studies are usually required before a permit is issued, especially if the waste is of a type that has not been handled before. Such studies can be very time-consuming, but knowledge of the subject is steadily increasing. These studies may well include:

- bioassays of the waste proposed for disposal;
- *in situ* bioaccumulation studies and dispersion studies;
- evaluation of the dispersion rate proposed in the area and its effect on the marine biology.

Dispersion studies relating to petrochemical wastes have in fact been reviewed in some depth.[8] It seems that the rate of dilution of the waste in seawater is mainly a function of the difference in specific gravity between the waste and the seawater. The dilution concentration at an arbitrarily selected time of one minute after dumping has been found to be largely a function of the rate of release, but it is also a function of the speed of the barge. Whilst such studies may contribute to a more efficient disposal of the waste in the ocean, ensuring that it is assimilated more quickly than it might otherwise be, the basic problems associated with this method of disposal are not being answered. Is it safe to dispose of waste in this way? What hidden damage, if any, is the result? There is at present no answer to such questions, yet it seems that the disposal of waste at sea, in the deep ocean, is going to continue.

References

1 Conway, R.A. and Ross, R.D., *Handbook of Industrial Waste Disposal*, Van Nostrand, New York, 1980.

2 Leenkeer, J.A., Malcolm, R.L. and White W.R., 'Investigation of the reactivity and fate of certain organic components of industrial waste after deep-well injection', *Environmental Science and Technology*, **10 (5)**, pp. 445–51.

3 Rich, L.A. 'Hazardous waste management – new rules are changing the game', *Chemical Week* (cover story), 20 August 1989, pp. 26+ (24 pages).

4 Boraiko, A.A. and Ward, F., 'Storing up trouble – hazardous waste', *National Geographic Magazine* (cover story), **16**, March 1985, pp. 308–51.

5 Winton, J.M. and Riach, L.A., 'Hazardous waste management – putting solutions into place', *Chemical Week*, 24 August 1988, pp. 26+ (15pp.).

6 Maugh, T.H. II, 'Incineration: deep wells gain new importance', *Science*, **204**, 15 June 1979, pp. 1188–90.

7 Kester, D.R., Burt, W.W., Cupuzzo, J.M., Park, P.K. and Ketchum, B.H. (eds), *Wastes in the Ocean*, **5**, 'Deep sea Waste Disposal' John Wiley, New York, 1985.

8 Ball, J., Reynolds, T.D. and Beckett, D.E., 'Deep sea ocean disposal of petrochemical wastes', Texas A & M University, College Station, Texas, 1980.

Chapter 8

LIVING WITH RADIOACTIVE WASTE

Radioactive waste is a fact which the world must learn to live with. Not all the radioactive waste that has to be dealt with and disposed of comes from nuclear weapons and nuclear power stations: a substantial volume comes from hospitals and a certain amount from industry which uses radioactive materials for a variety of purposes. As we have repeatedly pointed out whilst considering the various aspects of waste management, waste in general is one of the necessary evils of modern life, and radioactive waste is no exception. We have also stated as a basic principle that the safe disposal of waste is the responsibility of its creator: this is also true of radioactive waste.

Nuclear power arrived on the scene some 30 years ago – an awe-inspiring development full of promise for the future. When it first arrived on the scene, it was anticipated that its use would immeasurably enhance the quality of our daily life, providing energy in abundance at minimum cost. To quote the first chairman of the Atomic Energy Commission in the United States in those early, enthusiastic days: 'Nuclear energy ... would deliver electrical power too cheap to meter.' Unfortunately it has not worked out like that at all. The remark that we have just quoted, made by a senior and respected figure in the nuclear industry, displayed great optimism, but the industry has had to face many teething problems, of which waste disposal is but one.

Radioactivity: what is it?

The international unit for the measurement of radioactivity is the *sievert*. This is a unit defining a biologically effective dose of radiation. One sievert corresponds to an energy absorption of one joule per kilogram of tissue. A microsievert is a millionth of a sievert. What concerns us here is the fact that the impact of radioactivity on the human body *can* be measured. But what is not usually appreciated by the general public is the fact that we are all subject to natural radioactivity day in, day out. It is always there, coming from both the earth and the sky. The main contributors to natural radiation are cosmic rays, which come out of the sky, natural radionuclides present in the body, terrestrial gamma ray sources and the gas radon. The naturally occurring gas radon-222 comes from the decay of uranium-238, traces of which are present in the earth's crust. The gas emanates from the ground and appreciable concentrations can build up in dwellings, especially if they are poorly ventilated. Individual exposure, however, can vary considerably, the highest exposure being in regions of igneous rock such as granite. In such areas the exposure is around eight times the national average.

It is very easy to be 'alarmist' in this context. For instance, do you know that 360 million atoms of potassium-40 disintegrate in the human body every day? Nearly 2.4 million cosmic ray neutrons and about 10 million secondary cosmic ray particles penetrate the average individual every day! About 5,000 million gamma rays from radioactive sources in soil and building materials pass through that same human body every day! All this sounds horrific, but let us reduce it to terms that we can recognize. The biologically effective dose to the human body from sources such as cosmic rays, gamma rays and natural radionuclides totals perhaps 500 microsieverts annually, although some estimates put it as high as 2,000 microsieverts. It all depends upon what you eat, where you live and what you do. It is interesting to learn, for instance, that the amount of radiation we are likely to receive from natural sources is 20 times as great as that which we are ever likely to receive from artificial, man-made sources, such as a routine chest X-ray.

But what effect has this radiation on us? Let us begin by going to an extreme. It appears that, up to a certain point, damage to the body by radioactivity will be restored by natural healing processes, but if a person receives a rather high dose then that person will almost invariably die. However, of more immediate significance is the cancer-inducing effect of a low dose, and this is an area with a lot of uncertainties. This comes about because cancer is caused by so many things other than radiation and quite often the real cause cannot be determined. In Western Europe, cancer is one of the main causes of death, accounting for more than 25 per cent of all deaths. But very, very few of these deaths will be due to man-made radiation and hardly any to the existence of nuclear power stations.

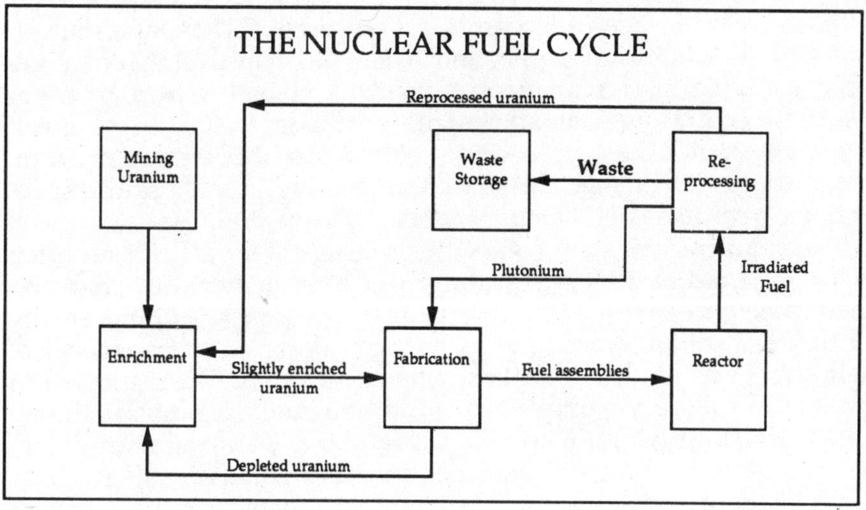

Figure 8.1 **The nuclear fuel cycle. This flow scheme illustrates the movement from raw material to waste processing and storage.**

Too dangerous to throw away – too valuable, too!

The nuclear industry complains that the public fuss and fear over the disposal of its radioactive waste products is out of all proportion to the real risks involved, and this is true, but that does not dispose of the problem. Radiation frightens because it is invisible, unfamiliar and little understood by the public at large. Nor have the mistakes made by those disposing of radioactive waste helped: the manner in which the nuclear waste processing plant at Sellafield in the UK has been operated, particularly over the early years of its operation, has brought the industry into disrepute. Yet radioactive waste is an inevitable by-product of nuclear energy, so the issue of its disposal must be faced. We use the term 'radioactive waste' in relation to waste arising from the initial mining of uranium ore through to the reprocessing or the long-term storage of the spent fuel, as set out in Figure 8.1. Initially it was thought that there would be a shortage of uranium ore, so steps were envisaged that would conserve its use. This introduces a reprocessing stage as indicated in Figure 8.1, but in practice the reprocessing of spent fuel has been successful in France only. The United States, the United Kingdom, Japan and West Germany all have reprocessing facilities and all except the United States plan to make use of them. However, at the present time (1988) reprocessing does not seem to be an economic proposition: the price of uranium ore is too low. The alternative to reprocessing is to store the spent fuel for reprocessing later.

The characteristic that makes this waste different from all other waste is its radioactivity. The atoms of certain elements are unstable: their nuclei 'decay' spontaneously. In the process they emit energy in the form of radiation – in particular, alpha particles (nuclei of helium), beta particles (fast-moving electrons or positrons) and gamma rays (very short-wave electromagnetic radiation). There are two significant points to be made about such radiation. First, it is dangerous. When it smashes into human tissue it gives its molecules an electric charge, thus ionizing them, which can cause havoc. The other significant fact is that this radioactivity is not permanent. It decays. Each radioactive element or isotope has what is called a 'half life': that is, half of its unstable (radioactive) atoms turn into other, stable forms within a certain time. This means, for instance, that after, say, 100 years the level of radioactivity remaining in used nuclear fuel coming from a nuclear reactor is only 1 per cent of what it was a year after its discharge from the reactor. After 10,000 years it will be 100,000 times less active than it then was, but even that level of radioactivity is roughly 500,000 times more active than fresh 'low-level' radioactive waste. This demonstrates the crucial difference between such spent fuel, which is described as 'high-level waste', and 'low-level waste'. Because of this, they require very different treatment for their safe disposal.

There is a great deal of literature on this subject and there have been many seminars organized to discuss the problems of radioactive waste disposal. Typical of these was the 1987 European Summer School on Radioactive Waste Management, held at Christ's College, Cambridge (UK) in July 1987. This approach to the problem is positive, but much of the treatment of this subject has been negative, although the fears raised are to some extent justified. Two investigative reporters gave a frightening, chilling account in a book entitled *Forevermore – Nuclear Waste in America,*[1] which documents the so-called scientific blunders, the criminal behaviour by industry, the cover-up by politicians and the impotence of the regulators over the years. The clean, trim and efficient nuclear power plants operating in the United States have produced radioactive waste that is now to be found in some 300 nuclear waste dumps scattered across the country. The original intention was to reprocess the spent fuel rods and recover the uranium and plutonium, but the three commercial plants built for this purpose have never been used. There is an inventory of what is stored in licensed dumps, but the fear is that other material may have been dumped illegally and some seems to have gone missing. Illustrative of the problem is the story concerning waste buried by the Atomic Energy Commission – plutonium waste so deadly that one ounce could kill 20,000 people. In 1974 it was buried in Maxey Flats, Kentucky with the assurance that it would move no more than an inch in 24,000 years: in 10 years it had moved hundreds of feet.

Government statements to the effect that radioactive waste posed little or no hazard and that the technology for reprocessing and disposal was proven, have been found to be untrue – so untrue that two fantastic methods of disposal, burying the waste in Antarctica and shooting it into space, both rejected 20 or more years ago, are once again receiving serious consideration. But of course, with such schemes the cure could be worse than the disease. Radioactive waste has been divided into three broad groups: low-, intermediate- and high-level. High-level waste generates a great heat, so that storage under ice could melt the ice cap and possibly raise ocean levels significantly. Launching it into space could cause widespread contamination since it may eventually come down.

Although all this sounds rather sensational, the underlying message is clear. Radioactive waste *is* highly dangerous and it should be disposed of properly. Unfortunately, in many cases management has acted irresponsibly, disregarding the dictum we have repeated so often: the responsibility for the disposal of waste must remain with its creator. The tragic story of Love Canal made it very clear that toxic waste from the chemical industry must be treated and rendered safe before disposal.[2] The same is true of waste from the nuclear industry, and the argument for proper treatment before disposal is all the stronger because the danger is hidden. Of course, the 'dumping' of radioactive waste, the usual means of disposal at present, merely stores up the problem for future generations to face and live with: surely that is not the answer?

What can be done?

So far, we have taken a look at some of the things that are being done but should not be done. Let us now be positive. What *can* and *must* be done? To answer that question we must understand the nature of the problem. Even the relevant data is so widely scattered that one author felt impelled to collect it and combine it in one volume.[3] All the relevant information is claimed to be included under five heads: physical data, chemical data, radioactive wastes, data for operations and general information. One review of this book tells us that only the conventional methods for managing radioactive waste are dealt with: the exotic options, such as disposal in the sea, sub-seabed, outer space and transmutation of nuclides are ignored.[4] Disposal at sea, the method that has been adopted by Sweden, as we discuss later in this chapter, may well prove to be the best practical answer: it is already being adopted by some countries despite violent opposition from the environmentalists. The root of the problem lies in the fact that there is no clear answer to the question: what level of radioactivity is 'safe'? As pointed out above, we are all subject to dosages of radiation all our lives from natural sources, so perhaps that level of dosage may be considered 'safe'. But not everyone would agree with that statement.

For disposal purposes, radioactive waste has been divided into three categories, depending upon the level of radioactivity. The nature of the waste falling under these three categories can be briefly defined as follows:

1 *Low-level.* This includes protective clothing, tools and the like. It is lightly contaminated with short-lived radioactivity. It represents no great danger, but there is a lot of it: 25,000 tons per year in the UK alone.
2 *Intermediate-level.* This includes reactor components, resins, and the like. It is fairly large in volume and is said to total some 35,000 cubic metres in the UK – enough to cover a football pitch to a depth of five metres – and is growing at the rate of 2,500 cubic metres per year.

3 *High-level.* The fuel from reactors, together with residual liquids after reprocessing. This is small in volume but presents the biggest headache, since it is highly radioactive, initially hot and much of the radioactivity is long-lived. It is this material that accounts for 90 per cent of all the radioactivity in the waste products from nuclear power plants. The present idea is to store it for 50 years, when it should be cool enough to be treated as intermediate-level waste.

To give some idea as to the proportions of these several types of waste, we present in Figure 8.2 the relative amounts of the three different types of radioactive waste that have accumulated in the UK. This data was originally presented with the vertical scale logarithmic, which we have also indicated to see how this data would then look. Whether a logarithmic scale was used to make the problem presented by the heat generating waste appear greater than it really was, we do not know, but it is in fact very, very small in terms of waste volume. For instance, the UK produces something like 50 million tonnes of coal mining spoil each year, which also presents a disposal problem, but it never reaches the headlines in the way that radioactive waste does. The low-level waste and some of the intermediate-level waste are already being disposed of: gases are released into the atmosphere and liquids are pumped into the sea, whilst the solid material is being buried on land or entombed in concrete and dumped into the sea. But, since 1983, both the UK and Japan have had to abandon that last method of disposal. The favoured option for low-level waste at the moment is to bury it 20 metres below ground in concrete bunkers set in clay or stable ground. Intermediate-level waste requires deeper burial, and the UK is using an anhydrite mine at Billingham, near Stockton-on-Tees. But these are short-term solutions and certainly do not cope with the problem in its entirety. All of the high-level and much of the intermediate-level waste is being stored on the surface until a final acceptable solution for its final disposal has been worked out.

It is very obvious that geology plays a vital role in preserving the integrity of hazardous radioactive waste stored in formations underground. A book written specifically for this purpose is therefore to be welcomed.[5]

Long-term disposal: the problem

There is no doubt that the long-term disposal of radioactive waste

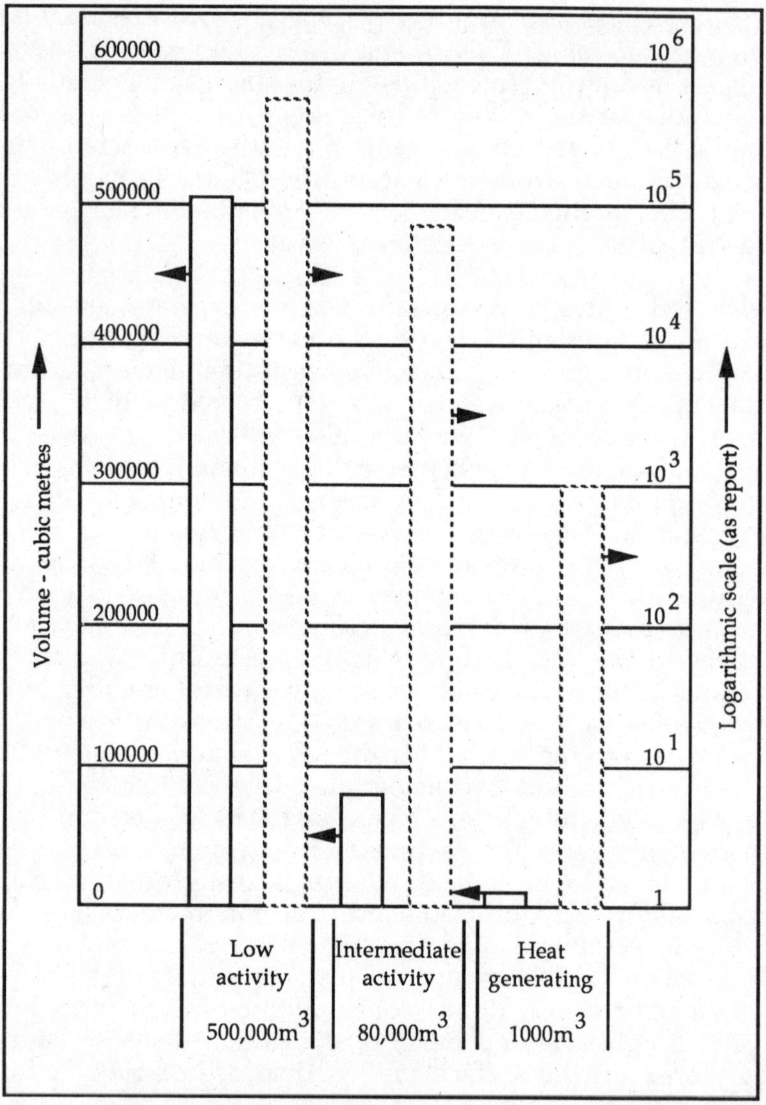

Figure 8.2 **Radioactive waste. This chart portrays the relative amounts of radioactive waste accumulated in the UK: it is estimated that by the year 2000 the low-level and intermediate-level waste will have doubled, whilst the high-level waste will have reached 4,000 tonnes. A more emotive way of stating that last increase would have been to say that it would have gone up 'four times', whilst the others have only doubled, but of course the total amount is still very manageable.**

remains a very serious problem. The issue has become politicized worldwide and the public attitude, urged on by the environmentalists and the so-called 'concerned scientists' has hardened. Long-term disposal relates especially to high-level waste, which is generally being stored in stainless steel double-walled tanks at the surface, where water is continuously circulated and cooled to keep the temperature down. This is expensive and there is always the risk of accident or, in today's political climate, sabotage. After some 10 years, the activity falls to a level at which the liquid can be concentrated, calcined and the residue processed. One suggestion is for it to be incorporated into blocks of glass. The proposed method for ultimate disposal of high-level radioactive waste is therefore:

1 *Solidification*. Let the waste cool. Then the radioactivity is much diminished and it can be solidified into borosilicate glass in a ceramic synthetic rock ('synroc').
2 *Packaging*. The synroc is then encased in special non-corrosive containers. Both inconel and concrete containers have been proposed.
3 *Backfilling*. Once the waste has been placed in the depository, there should be buffer materials, such as bentonite clay, around the containers, both to keep them apart and to bury them.

So far, so good. The highly radioactive short half-life materials will decay to levels comparable to those already obtaining in the earth's crust. But what about those waste products where the radioactivity is long-lived? It is said that to ensure ultimate safety for such products, the repository should be in a very stable geological formation, well away from earthquake zones, never likely to be disturbed either by property developers or prospectors for valuable deposits, either now or even in 1,000 years' time. The site should also be free from incursion by ground water, which might possibly carry some of the radioactive material away. Such formations can be found in granite and basalt structures, salt domes and some clay strata. But no method of disposal can be 100 per cent safe, and the whole purpose of the elaborate procedure we have just described is to minimize risk. But this is to consider that risk in isolation. The risk arising from radioactive waste disposal should in fact be set against the risks that would come if there was no nuclear power. If the nuclear power plants were shut down, then no more such waste would be created by that industry, but instead there would be the risk occasioned by using coal, with its consequent sulphur and carbon dioxide emissions, which bring acid rain and may eventually lead to a major shift in the climatic belt. The risk from coal-fired power stations, although far higher than that from nuclear power stations, has been endured for decades without much protest. Thus far the problem. Is there a solution?

Long-term disposal: the solution

It appears that there *is* an acceptable solution to the disposal of radioactive waste. On 24 October 1985 Passant, then head of the UK Central Electricity Generating Board's nuclear waste technology gave a well-attended lecture on radioactive waste disposal.[6] He put the subject into context by telling his audience that £250 million was going to be spent at Sellafield to reduce the discharges from that plant 'quite drastically' over the next few years. The end result would be to save 'potentially' (there was no certainty) two lives over 1,000 years. He then invited his audience to ponder how many 'real lives' might be saved were such a sum to be spent in some other way. He made the point that the usual categorization of such wastes as low-, intermediate- and high-level, depending upon their radioactivity concentration, is an oversimplification. The boundaries are actually hazy, but of course where doubt existed the higher category was always chosen, to be both conservative and extra-safe. He asserted that shallow burial was completely adequate for the short-lived intermediate waste when sealed in steel or concrete containers, the waste within having first been solidified. A vitrification plant is being built at Sellafield for this purpose using a proven French process. Hence, in his view, the problem of radioactive waste disposal is now solved for the time being. Eventually a home will have to be found for the high-level waste, but there is no urgency about that: its volume presents no problem. At the moment, in the UK, it is all stored in cooled tanks at the British Nuclear Fuels plant at Sellafield, awaiting vitrification. Then it will have to be stored somewhere. Both of these are expensive steps, but have been reflected in the cost of electricity to the consumer.[7] It is in fact a valuable resource, containing radioactive isotopes which have both industrial and medical uses. In addition, there are the rare but precious platinum metals – rhodium, palladium and ruthenium – and the chemical separation of these valuable components is both technically feasible and economically attractive.

Radioactive waste disposal worldwide

So far we have been looking at the problem of radioactive waste disposal in the UK, which deals not only with its own waste but also with the waste from other countries, particularly Japan. When we turn to look at the United States we see a much greater problem, although it was not created by the nuclear power industry but by the defence industry. For instance, at the Savannah River Tank Farm near Aiken in South Carolina some 33 million gallons of high-level radioactive waste from military sources have been stored in steel tanks for some 40 years. This volume is some 70 per cent of the total stored in the United States.[8] A clean-up operation was started some three years ago and is likely to cost more than US$1 billion. The political reaction is interesting. Apparently most of this waste is created in the eastern United States but may be dumped in the west, the Department of Energy having dropped plans to look for suitable sites in the originating area.[9] There is, therefore, a strong political campaign, with senators seeking to halt the funding of the clean-up process, in order to impose a year-long pause to permit a careful review.

Of all the radioactive wastes in the United States, those posing the most serious risks are located at the Hanford site in Washington State, run by Westinghouse, and the Savannah River plant, run by DuPont. There is clear evidence of soil and water pollution at Savannah River and, according to DuPont, the severe contamination from decades of dumping 'may extend well beyond the period for which land control can be anticipated and indeed may exist for centuries or millennia'.[10] So the problem is very serious indeed.

Several other countries, notably France, South Korea, The Netherlands and Taiwan are considering an offshore disposal system known as POWER (Pipeline Operated Waste Energy Repository). This concept is being marketed by Warrior Resources who say that it is cost-effective when compared with the Westinghouse Surepak scheme for the burial of low-level waste and the shallow land burial of such waste. In addition, it would eliminate the need for a separate deep repository for intermediate-level waste.[11]

Whilst Japan has been sending its high-level waste to the UK and France for processing, it wishes to be self-reliant in this respect. Thus, the Power Reactor and Nuclear Fuels Corporation (PNC) plan to build a storage facility at Hornobe on the northern Japanese island of Hokkaido. This island has 20 per cent of the land area of Japan but is sparsely populated, having only 5 per cent of the population, but nevertheless the company is meeting strong resistance. The present contracts with the UK and France stipulate that the solidified waste is Japan's property, but the situation is seen by the Japanese as 'living in a house without a toilet'.[12] Meanwhile, high-level waste is accumulating in Japan at the rate of 700 tonnes per year, since they now have some 30 nuclear power plants.

Sweden has yet another solution.[13] By 1990 they expect to have what has been described as a 'Rolls Royce' solution, known as SFR. It will cost US$200 million and is located 50 metres under the seabed, near Sweden's Forsmark nuclear station. Sinking the material under the seabed ensures that the water around it is stagnant, an important factor because water movement is the main mechanism for transferring radioactivity. The Swedes say that SFR, which consists of concrete silos and chambers, will operate safely for at least 1,000 years. In half that time the waste should have decayed to a harmless level. The process of finding suitable sites is also in hand in Germany, France and the United States. Safety is, of course, the primary consideration, and mathematical models are being used to ensure this.

The media may well have sensationalized this issue by the use of such emotive phrases as 'hidden horrors' and 'nightmarish contents'. Writing about a waste processing plant being built at Sellafield they say: 'A vitrification plant is being built at Sellafield which – if it works – will glassify the hot liquid waste and turn it into more easily manageable glass blocks'. But why the qualification 'if it works'? The process is already working very satisfactorily in France, so we have here yet another example of the way in which the media continually seeks to dramatize the problem and frighten the public. Nuclear waste *is* dangerous, but it can still be disposed of quite safely.

But questions remain ...

Just to get a 'feel' for the problem radioactive waste poses in the United States, it appears that some 2,000 tonnes per year of 'hot' waste, largely plutonium, is accumulating from power and weapons plants. The Yucca mountain in Nevada is being considered as a burial site for this 'most poisonous issue'. Its suitability is being studied and, if the Nuclear Regulatory Commission approve, construction work may start in 1998. Five years later canisters filled with this deadly waste would be placed there, to a total of 70,000 tons. Some 50 years from now the site would be full, and it is assumed that, in another 300 to 1,000 years, 'nature would have done its job'. The site is remote, about 20 km from the nearest settlement, and comparatively dry.[14] But there are many voices being raised, expressing concern: such long-range predictions *must* be suspect. The statement: 'You're stretching science to and beyond its limits to project the [radiation] doses tens of thousands of years into the future' is typical.[15] Nevertheless, the immediate problem is also receiving attention. A paper by Heafield and Barlow reviews the fundamental objectives of radioactive waste management and demonstrates their application to selected waste management processes.[16]

What next?

A two-day conference in London (25–26 February 1988) debated the theme 'Radioactive waste management – what next?' in the light of the concern being expressed, and the continuing pressure coming from both local and international public opinion. Thirteen papers were presented on various aspects of this subject by speakers drawn from government, the nuclear industry, private industry and the environmentalists.[17] The scope and scale of the problems facing the nuclear industry with respect to radioactive waste disposal is well illustrated by the titles of some of the papers:

- Minimization of waste in process design
- Determining the tolerable risk
- Effect of planning law and the environmental impact assessment directive
- Deep disposal options for low- and intermediate-level radioactive waste
- Disposal under the seabed
- The technical requirements of environmental impact assessment for radioactive waste repositories
- Resource consequences of reducing disposal of radioactive waste to the environment
- Factors which can affect present and predicted radioactive waste arising in the UK
- Should the government be involved in radioactive waste management strategy?

Here are concerns and phrases that we have met many times

before: only the subject is different. Will the industry be successful in finding an acceptable solution to an ever-growing problem? Meanwhile an in-depth research study, the details of which have been advised to us by Dr. A. Makhijaney, recommends that high-level radioactive waste be stored for about 100years, until a staisfactory long-term solution can be found.

References

1 Bartlett, D.L. and Steele, J.B., *Forevermore – Nuclear Waste in America*, W.W. Norton (USA), 1985; see also 'Perils of the nuclear trash can', review of the book, *The Economist*, **297**, 19 October 1985, pp. 107–8.

2 Kharbanda, O.P. and Stallworthy. E.A., *Safety in the Chemical Industry: Lessons from Major Disasters*, Heinemann, London, 1988.

3 Stewart, D.C., *Data for Radioactive Waste Management and Nuclear Applications*, Wiley, Chichester, 1985.

4 Aston, S.R., *Jnl. Environmental Radioactivity*, (review of book, ref. 3) **3**, 1986, pp. 315–7.

5 Krauskopf, K.B., *Radioactive Waste Disposal and Geology*, Chapman & Hall, UK, 1988.

6 Report: Prospects for waste disposal. *Atom 352*, February 1986, pp. 28–31.

7 Thornback, J., 'Radioactive waste – a valuable resource', *New Scientist*, **108**, 14 November 1985, pp. 76–7.

8 'The buried cost of the Savannah River plant', *Science*, **233**, 8 August 1986, pp. 613–5.

9 'Power-full solution to the waste problem', *Nuclear Engineering International*, July 1986, p. 11.

10 Alvarez, R. and Makhijani, A., 'Hidden legacy of the arms race – radioactive waste', *Technology Review*, **91**, August/September 1988, pp. 42–51.

11 'Power-full solution...', *Nuclear Engineering International*, op. cit.

12 Johnson, B., 'A Country without a "nuclear toilet" ', *New Scientist*, **108**, 3 October 1985, p. 25.

13 Highfield, R., 'Where will they put it now?', *The Daily Telegraph*, 2 May 1986, p. 9.

14 Alvarez and Makhijana, op. cit.

15 Horgan, J., 'A safe site?', *Scientific American*, **258**, June 1988, p. 18.

16 Heafield, W. and Barlow, P., 'Management of wastes from the nuclear fuel cycle', *Nuclear Energy*, **27**, December 1988, pp. 367–76.

17 Proceedings: 'Radioactive waste management – what next?', London, 25–6 February 1988, IBC Technical Services Ltd., UK.

Part III
THE REAL SOLUTION

Chapter 9

HAZARDOUS WASTE MANAGEMENT

Most hazardous waste originates from industrial activity of one sort or another. Hospitals are a minor source of hazardous waste, and there will be other minor sources, but all these we will disregard. We propose to concern ourselves with hazardous waste in all its aspects, including the technical developments and the legal aspects of control in various countries. Radioactive waste is currently receiving a great deal of public attention, since it presents certain special hazards, and we have therefore dealt with this separately in Chapter 8.

A problem that cannot be buried

This is the title of a typical article in the popular press on the subject of hazardous waste, with the alarming but perhaps true subtitle: 'The poisoning of America continues'.[1] A selection of the titles appearing over the multitude of articles that have been, and are, appearing in both the technical and the popular press on the subject of hazardous waste gives, we feel, a flavour of the subject:

- 'The elusive pursuit of toxics management'[2]
- 'Who gets the garbage? – In the Third World, a new sense of alarm about toxic waste'[3]
- 'The heat is on – chemical wastes spewed into the air threaten the earth's climate'[4]
- 'Management of hazardous and toxic wastes'[5]
- 'Hazardous waste management – new rules are changing the game'[6]
- 'Who will clean up by cleaning up?'[7]

- 'The missing links: restructuring hazardous waste controls in America'[8]
- 'Living dangerously with toxic wastes – three tormented towns point up past, present and potential problems'[9]
- Problem of toxic emissions – officials find lack of data a hindrance'[10]

The above headlines present a graphic picture of the fears and the inherent dangers of toxic waste. Clearly the picture is a grim one. These headlines relate largely to one of the most industrially advanced countries in the world, but the position in the rest of the world is no better: in all likelihood it is much worse. In some respects, the situation in relation to the developing world is pathetic, with either no laws whatever governing the handling of toxic wastes, or with laws that are not – indeed, cannot – be enforced. To some extent it was this type of laxity that led to the terrible disaster at Bhopal, when more than 2,000 died. However, industry worldwide, and particularly the chemical industry, has reacted wisely following that dreadful event at Bhopal during the night of 2 December 1984. The chemical industry has now become a yet safer place of work, and the communities living close to its factories feel somewhat safer. However, whilst industry has learnt much from this particular accident, that learning process is, and must be, ongoing. Accidents will happen, but each accident brings its lessons, if we will but listen and learn.

Lack of data is a problem

The last of the headlines quoted above raises an issue of great significance in relation to waste management, and more particularly, toxic waste management.[11] The article is based on a 1985 Congressional Survey of chemical emissions, to which 303 chemical companies responded. There was also another exhaustive survey of 86 major chemical companies conducted in the same year by the House subcommittee on health and the environment – all this, of course, in the United States. The *New York Times* analysed this data and the findings, and further interviewed the executives of a number of chemical companies, together with individuals in the regulatory organizations, environmentalists and others concerned with the subject. Some of the most significant but somewhat shocking conclusions were:

1 There are no specific regulations governing the emission of

many toxic chemicals, at least in some states.
2 Companies have widely differing attitudes to this subject.
3 Most emissions come from a few large plants, perhaps less than ten per cent of all chemical plants.
4 Old plants emit far more toxic chemicals than the newer plants.

It seems that the regulation of such matters is a state, rather than a federal, responsibility and this has resulted in some strange paradoxes. For example, a chemical factory in Parkersburg, WVa. was said to be emitting 500,000 kilograms of acrylonitrile, a known carcinogen, into the air every year without violating any existing state law. The same emission in Philadelphia or New York would be liable to prosecution, since it far exceeds the stipulated permissible maximum in those cities.

This confusion and disparity is also to be seen amongst the member countries of the EEC. A member of the Commission of the European Communities, writing to the journal *The Economist* (**308**, 30 July 1988, p. 4) whilst recognising that the trade in toxic waste has often been conducted with complete indifference to the potential dangers to the environment and human health, makes the point that the European Commission has been concerned about this issue for some time and is taking action. Not only are member states being pressed to adopt existing legislation, but new legislation is being drafted that seeks to ensure that recent scandals will not recur. The type of scandal that is in mind is highlighted by a headline 'Be my cesspit'.[12] It appears that the Nigerian government has told the EEC that it is suspending full consular relations with Italy because Italian companies have dumped dangerous waste in Nigeria in breach of community law. The directive in question sets out strict conditions for the export of toxic waste, one being that the agreement of the recipient government should be secured. The trouble lies in the fact that though this directive was issued in January 1987, a number of countries, including Britain and Italy, have yet to incorporate the directive into their national law. So the Italian exporters were within their own country's legal requirements, although acting in defiance of the EEC directive. The EEC countries are not the only ones seeking to dispose of their waste on the continent of Africa: Table 9.1 lists some of the major incidents, showing that the problem is worldwide.[13] Unfortunately, such waste disposal can offer substantial financial inducements to poor countries.

Table 9.1 Major disposal of waste in Africa

Country	Nature of waste disposed of
Angola	Has taken toxic waste for disposal from Europe.
Benin	The government has agreed to take 5 million tonnes of industrial waste each year from North America and Europe.
Congo	A deal to dump one million tonnes of chemical waste from the US and Europe was cancelled in 1988.
Gabon	Has agreed to take radioactive waste from uranium mining in Colorado.
Guinea	15,000 tonnes of toxic flyash was dumped on Kassa Island before being returned to the US.
Guinea-Bassau	A five-year contract to take 15 million tonnes of pharmaceutical and tanning wastes was recently cancelled.
Liberia	Proposes to import hazardous waste, including contaminated earth from West Germany.
Nigeria	4,000 tonnes of Italian waste dumped at Koko before being returned to Italy.
Sierra Leone	Has accepted American toxic ash containing cadmium and mercury.
South Africa	120 drums of waste from New Jersey containing sludge laced with mercury was dumped
Zimbabwe	7,500 litres of hazardous waste from the US Armed forces was tipped into a phosphate tip.

Adapted from *The waste trail leads south*, R. Milne, *New Scientist*, 1 April 1989, p. 25.

Waste management is big business

Waste management, and more particularly toxic waste management, whilst it poses a tremendous challenge and a continuing threat, nevertheless presents some significant opportunities. It is 'big business' these days, especially for a company that can open up a new field. It has been estimated that expenditure on waste management in the United States alone will be more than US$300 billion over the next 50 years.[14] This 'carrot' is attracting many substantial, and even conservative, companies into the waste management business, despite the fact that it deals with toxic, highly hazardous materials, and is full of economic, legal, political and even technological problems. International construction giants, such as the Bechtel Group, now have a significant stake in the business of waste management, whilst companies who have established a 'niche' for themselves in this ever-growing market take full advantage of it. For instance, it seems that the cost of using dumps, particularly those with permits that allow them to accept hazardous waste, is on the 'up and up': they are an increasingly scarce commodity.

Companies operating in this field, and offering waste disposal services and the related technology are expecting a sales growth of between 20–30 per cent per year, and this is reflected in the price of their shares. The business has seen several poor years but now that industry is thriving, the shares of some waste management companies are doing very well indeed. For instance, the shares of Kidder, Peabody & Co. rose by nearly 200 per cent in 1985. This should be compared with an average rise of around 30 per cent for industrial companies in general, as reflected in Standard & Poor's 500-stocks index. Facts and figures are hard to come by in this field, but it is estimated that private industry has been consistently spending far more than the government on the cleaning up of waste, particularly hazardous waste. One indication of this is the progress of Arthur D. Little, a major consulting firm based at Cambridge, Massachusetts. This firm has expanded its hazardous waste group from a mere six to nearly 40 professionals, but is still unable to cope with all the enquiries that come flooding in. Many have to remain unanswered.

The subject is also dealt with at a multitude of seminars and symposia, typical of which are:

- *Chemical Week/Arthur D Little*: The 'ninth executive briefing' on 'Today's business opportunities in hazardous waste' is announced, to be held in New York City in June 1988. The brochure carries an illustration of a drum, presumed to hold hazardous waste, and a graph showing the extent of the business available in this area. It has risen from US$1.0 billion in 1981 to US$2.0 billion in 1986, and is forecast to rise to US$7.5 billion by 1991.

- *Chemical Equipment*: This trade journal, together with several other co-sponsors, announces *HazMat 88*, a conference and exhibition dealing with hazardous materials management. It was claimed to be the 'largest, most productive conference and exhibition in the hazardous materials management field... to update attendees on the latest developments and technology'.

It seems that a host of companies specializing in the hazardous waste management field have sprung up and are making bold claims. The following, culled from their advertisments, are typical of these:

> We'll manage your site from start to finish with the assurance of successful closure.... We've got the experts to integrate innovative and proven technologies to produce solutions which minimise your costs.

> [We are] a full service company uniquely positioned to provide total management... [we] simplify the administrative process tied to complex project management.

> Industrial and hazardous waste destroyed... effectively and efficiently... incineration and energy recovery systems based on technology proven at installations around the world.

> The nation's largest commercial capacity... [we] have *permanent* solutions... just pick up the phone, we'll pick up your problem.

But let us assess all this in context. The fact that the waste management business is growing rapidly does not necessarily mean that there are easy profits to be made. The scenario is calling for rapid transformation. Once a slovenly, ill-controlled and supervised business, it has now become a sophisticated and well-run industry, which will show a profit only if it is operated extremely efficiently. In the United States alone, there is a legacy of several thousand problem sites, of which perhaps a dozen have been 'corrected' over the past few years. It is interesting to note the change in terms: sites are now 'corrected' rather than 'cleaned up'. It has been found impossible to completely clean up badly managed dumps: they can only be made less dangerous. With this history of mismanagement behind them, the companies now in the business are obviously going to be constantly under suspicion. Those most likely to benefit in the short term are the major garbage collection firms, especially those who have licensed landfill areas at their disposal. But whilst the business can be profitable, it is also at risk: there is likely to be a spate of liability suits at the slightest violation, whether intentional of accidental.

Poisoning both ourselves and posterity

Evidence has been steadily accumulating for a number of years now that unless the chemical wastes that have been released without proper treatment for so long are properly dealt with, they constitute not only an immediate and very serious danger, but also a long-term danger, not only to posterity, but to planet earth itself. This is by no means an exaggerated or alarmist view. It seems that apart from the direct effect that toxic chemicals have on the environment, there are long-term effects due to ozone depletion by the man-made chlorofluorocarbons that have invaded the upper atmosphere, and the ever-increasing percentage of carbon dioxide in the atmosphere due to the burning of fossil fuels, bringing with it what has been called the 'greenhouse effect'. The magazine *Time* reports that already a satellite image recorded in October 1987 indicates a significant 'ozone hole' over Antarctica.[15] Because of aberrations such as these, it may well be that the earth's climate is no longer stable, with the result that drastic adverse and unexpected changes in the climate may well occur. Who knows? What is certain is that the scientists are not sure, so that the only sensible course is to minimize the emission of toxic and potentially harmful wastes to the maximum extent possible. Anything less is but to invite trouble and be suicidal.

In this context it would seem that the disaster at Bhopal had a beneficial effect, since not only did it lead to a spate of safety audits by chemical companies across the world, but also the operations of such companies were assessed in context. In other words, their impact on their local communities was carefully studied in the specific context of safety. An answer was sought to the question: 'what threat, if any, do they present to the local community?' As a result the chemical industry is spending ever-increasing amounts on cleaning up its operations. Indeed, it is estimated that, in the United States alone, the amount that will be spent during 1988 on environmental and pollution control equipment is likely to be US$3 billion. This represents nearly

17 per cent of the industry's total capital expenditure, and means that one dollar in every six will be invested in plant and equipment that brings no direct return. Nevertheless, such a policy is essential. Strict guidelines now exist with respect to the reduction and disposal of hazardous waste in the United States, some 75 per cent of which is believed to be produced by the chemical and process industries. Compliance with the regulations is going to cost a great deal of money: not only has more to be done than was done in the past, but the cost of doing it is rising fast – several-fold in the past few years. Some attempts have been made to recover such costs from insurers. In a test case, Shell Oil took nearly 300 of its insurers to court in order to recover some US$3 billion that the company expects to have to spend on cleaning up its waste at just one site – Rocky Mountains Arsenal, Denver. Failing this, the extra costs will ultimately have to be borne by the consumer and the prices of a very wide range of products will therefore be going up.

It is very clear that toxic wastes just cannot be buried and forgotten, either literally or figuratively. The 'nettle' really will have to be grasped. Hardly a day goes by without a report appearing in the media where a community has discovered that they are living on, or next door to, some deadly poison: sometimes a horrid brew. The names of some of these noxious, toxic chemicals have become notorious: they are now household words. We hear of dioxins, vinyl chloride, PBBs and PCBs, together with some of the more old-fashioned 'heavy metals', such as arsenic, lead and mercury.[16] What is more, the problem is getting worse by the day. The number of problem sites is continually increasing and it has proved very difficult to clean them up. If, once again, we take the data for the United States as an example, it is believed that there are at least 10,000 hazardous waste sites which may well cost US$100 billion to clean up, or over US$1,000 per household. In addition, it is said, there are perhaps 400,000 waste tips and dumps which, whilst not hazardous in themselves, need some sort of corrective action to render them innocuous. This is but one country: what then is the situation worldwide?

That the problem is indeed worldwide is demonstrated by the visible effects that are so well known to us all. We have acid rain, smoggy skies, and foul beaches. The danger, whilst at first hidden, eventually makes itself apparent. Typical of the incidents that reach the media is the story of the seals. The sea is a great 'sink' receiving millions of tonnes of industrial waste: one result – thousands of dead seals.[17] Even though the link between pollution and the dead seals has not been firmly established, such a link probably exists. Another hidden danger is toxic materials stored in steel drums which are eventually released as the drums rust away. Then the poisons seep into the earth and represent an almost irreversible threat to water supplies, public health and eventually the national economy as a whole. The problem stays with us in part because governments fail to act decisively. In the United States, for instance, the EPA seems confused and ineffective. Only about 25 per cent of the money that has been sanctioned for clean-ups has so far been spent, and it is feared that, at the current pace of work, 'millions of Americans will wait decades for the EPA to clean up their poisoned communities'.[18]

The Third World awakens

Nevertheless, as we have said, the problem is by no means limited to the United States. It is a worldwide problem. There is now a new sense of alarm about toxic wastes in the Third World. Their problem is twofold. First, there is the threat of the waste materials they are generating within their own borders as industrialization progresses; second, they have the additional problem of toxic waste deliberately imported from other countries and dumped within their borders. Apparently, unlike the trade in drugs, ivory or slaves, dumping toxic waste and poison in poor countries seems to contravene no existing law. Whilst the situation in relation to legislation, regulation and inspection in these countries is lax, the prime motive is financial. The source is usually a developed country, and it is far cheaper to ship such materials out than to dispose of it properly at home. The country accepting such materials earns scarce and valuable foreign currency which will then pay for essential imports. The following represent some typical stories, illustrating the sort of thing that happens:[19]

Trouble in Turkey

A West German waste transporter company persuaded a Turkish cement plant to accept 1,500 tonnes of waste-laden sawdust for use as fuel. But this material, instead of being burnt, was left in the open for several months, until a local newspaper reported that the material contained PCBs. The company was then forced to bring the material back home under strong pressure from both the Turkish and West German governments. The consequences, had the material in fact been burnt, could well have been horrific.

Storing 'safe' chemicals in Nigeria

Early in 1988 an Italian businessman persuaded a retired Nigerian timberworker to store more than 8,000 drums of certified safe chemicals in his yard in the town of Koko, in south-western Nigeria, for a monthly rental of US$100. It appeared to be a perfectly innocent deal until several of the drums started to leak, spilling a foul-smelling liquid on the ground. This was found to contain the deadly polychlorinated biphenyl (PCB), one of the most toxic chemicals known. The documents declaring the contents were manifestly false. The Nigerian government moved swiftly, recalling their ambassador from Rome and detaining both an Italian and a Danish freighter thought to be amongst the five vessels that had brought the deadly cargo into Nigeria. The Italian businessman, meanwhile, had disappeared.

A problem with incinerator ash

Incinerator ash can well be harmless, and has indeed been used for making breeze blocks. But the 15,000 tonnes that were dumped on the uninhabited island of Kassa, just off the coast of the West African nation of Guinea, was found to contain deadly cyanide, with lead and chrome byproducts. It was brought in by a Norwegian shipping firm: its local manager was also Norway's honorary Consul-General there. He was detained, but later released on the assurance that the toxic ash would be removed on a Norwegian ship.

Singapore finds a haven for its deadly waste

Companies in Singapore had apparently been secretly sending some of their chemical industrial waste to neighbouring Thailand for years – since 1981, in fact. However, some 600 drums of this waste was misrouted and languished in a dock at Bangkok where it was ignored until a mysterious yellow, toxic liquid started oozing from the drums. Apparently the misrouting was deliberate. The drums had been shipped to bogus Thailand firms, and so had sat there unclaimed for years. Officialdom, bound up with red tape, had never got round to action.

These stories from across the world are apparently typical of what is happening all the time, and will continue to happen until legislation and the related procedures in a number of countries are tightened up.

The missing links and the changing rules

There is no doubt that the appropriate technology is available for the proper treatment of hazardous industrial waste. It *can* be rendered harmless – sometimes cheaply, sometimes at a cost – but it *can* be done. However, for the available technology to be properly deployed, a new managerial approach is required.[20] Cost must no longer be the prime and dominating consideration and, to some extent at least, the profit motive must be subordinated to the common good. The stories related above, which involve companies operating in highly industrialized societies, permit us to see what can happen when the profit motive dominates, but the governmental regulating agencies, such as the EPA, also have a role to play. It is essential that they cooperate with both industry and the community, helping forward the selection of the most appropriate way of managing the wastes arising in any given situation. It is cooperation, not confrontation, that is required. It seems that Europe is providing a lead in this regard, a lead which other countries, including the United States, could well learn from. Several European countries have now evolved sound and very effective regulatory and disposal systems for the treatment of hazardous waste, which provide a much more sensible solution to the problem than that of landfill, still so widely used in the United States.

Typical of this new approach is that adopted by Denmark in 1973. The government established what is called the Kommunekemi facility at Nyborg – a plant designed to destroy more than 90 per cent of Denmark's hazardous industrial waste by incineration. This integrated facility also recovers sufficient heat from the toxic waste incinerators to supply Nyborg's 18,000 inhabitants with about one-third of their total heating requirements. The West German government has adopted a similar approach, coordinating the construction of 15 treatment plants, designed to destroy most of the toxic industrial waste produced in that country. Several other European countries, notably Austria, Finland, The Netherlands and Sweden, are following these early examples and putting ever greater emphasis on incineration, treatment and recycling as the means of disposal. The age-old practice of dumping, still so prevalent in the United States, has now been largely abandoned in Europe. How has this come about? There are a number of reasons, the chief of which seem to be:

- The European countries have, in general, socialistic institutions, in sharp contrast to the free enterprise tradition that prevails in the United States. Further, they have a long history of centralized decision-making, which inhibits the legal battles so prevalent in that country.

- Comprehensive regulatory reforms have proved to be more effective than the offering of financial incentives to private enterprise.
- A system which involves cooperative financing and management between both regional government and the private industry in their area has been very effective.
- In Europe there is wide public distrust of disposal via landfill, which has led to its replacement by viable treatment technologies.

Whilst the hazardous waste management industry is growing fast, as more and more information becomes available on the adverse effects of toxic waste, the laws and regulations are becoming ever stricter. This means that, although the industry strives to meet the requirements with the appropriate disposal strategies and management systems, the rules are actually changing in the 'middle of the game'. Until the last rule is in place, no one can be really sure whether they are coping properly and effectively. However, it seems that that position will never be reached, since new information is always coming in, precluding the prospect of any finality.[21]

It is an elusive pursuit

The frontiers of technology are steadily advancing, and this is particularly true in relation to toxic waste management. So much work is being done in this field, and the interest is so intense, that the first international conference on recent advances in this area was held in Vienna in 1987.[22] One would have thought that this growing knowledge would have led to a growing mastery of the situation, but this does not seem to be the case. In part, this is due to the 'sins' of the past. Each new discovery of an abandoned hazardous waste dump, such as those at Love Canal and Swartz Creek, discussed earlier, gives rise to public fear verging on chemophobia.[23] Not only is it very difficult and very expensive to deal with such dumps effectively, but each new leaking underground storage tank, every spill on the highway, any release of toxic contaminants from a factory chimney, immediately makes headlines in the media. This further fuels the public response, manifested in a growing antagonism to everything 'chemical'. A helpful term devised to select the most appropriate technology is BPEO (Basic Practicable Environmental Option). It is defined as:

> ...that alternative which achieves the minimum adverse impact on targets in all media in the long term as well as the short term, at practicable costs.

But that is not always as easy as it sounds, since such a selection demands judgement and is also very dependent upon local circumstances.

It has now become apparent that it is simply not good enough just to get toxic waste 'out of the way', especially where this means 'storing up trouble'. What is required is effective waste treatment and proper management, so that the problem is solved, not merely postponed. It seems that toxic waste policy-making has, or is going through, the following four distinct phases:

1 optimism about the conventional market approach;
2 response to public frustration with traditional policy-making;
3 acceptance of broader equity-based approaches to the problem;
4 adoption of an integrated approach to waste management.

When we look at the United States , it is still struggling through phase 3, but a number of European countries are well into phase 4. Much of the developing world is probably still stuck at phase 1. Of course, the integrated approach must be the ultimate goal since it takes into consideration all sources of hazardous emissions and ensures their proper control. This calls for team effort: industry and government agencies must cooperate and complement each other for the best and most effective results. This demands new levels of mutual trust and respect in many countries – trust and respect which has to be earned. Whilst broad policy goals have to be mutually agreed, together with ways and means, the specific choice of a disposal strategy has to be left to industry itself. The companies producing the waste know most about their operations and about the hazards it presents. Despite the progress that has been made, many formidable problems remain, and a great deal of innovation and creativity will be required to solve them. But much can, and should, be done *now* by a wider adoption of the means already being successfully used in some countries.

References

1 Magnuson, E., 'A problem that cannot be buried', *Time*, 14 October 1985, p. 24+ (5 pp.).
2 Mazmanian, D. and Morrell, D., 'The elusive pursuit of toxics management', *Public Interest*, Winter 1988, pp. 81–98.
3 'Who gets the garbage? – in the Third World, a new sense of alarm about toxic waste', *Time*, 4 July 1988, pp. 24–5.
4 Lemonick, M.D., 'The heat is on – chemical wastes spewed into the air threaten the earth's climate', *Time*, 19 October 1987, pp. 62–9.
5 Crittenden, B. and Kolaczkowski, S., 'Management of hazardous and toxic wastes', *Process Engineering*, May 1987, pp. 81–3.
6 'Hazardous waste management – new rules are changing the game', *Chemical Week*, 20 August 1986, p. 25+ (26 pp.).
7 Main, J., 'Who will clean up by cleaning up', *Fortune*, **113**, 17 March 1986, pp. 96–102.
8 Piasecki, B. and Gravander, J., 'The missing links: restructuring hazardous waste controls in America', *Technology Review*, **88**, October 1985, pp. 41–52.
9 'Living dangerously with toxic wastes – three tormented towns point up past, present and potential problems', *Time*, 14 October 1985, p. 29+ (5 pp.).
10 Diamond, S., 'Problem of toxic emissions – officials find lack of data a hindrance', *New York Times*, 20 May 1985, pp. D1, D5.
11 Ibid.
12 'Be my cesspit', *The Economist*, **308**, 16 July 1988, p 146.
13 Milne, R., 'The waste trail leads south', *New Scientist*, 1 April 1989, p 25.
14 Main, op. cit.
15 Lemonick, op. cit.
16 'Living dangerously with toxic wastes…', *Time*, op. cit.
17 'North Sea pollution – the 7000 seals', *Economist*, **308**, 27 August 1988, p. 78.
18 'Living dangerously with toxic wastes…' *Time*, op. cit.
19 Anderson, H., 'The global poison trade – how toxic waste is dumped on the Third World', *Newsweek*, 7 November 1988, pp. 8–11; also 'Who gets the garbage?…', *Time*, op. cit.
20 Piasecki and Gravender,op. cit.
21 'Hazardous waste management…', *Chemical Week*, op. cit.
22 Crittenden and Kolaczkowski, op. cit.
23 Mazmanian and Morell, op. cit.

Chapter 10

MINIMIZATION OF WASTE BY PROCESS DESIGN

Waste is the ever-present curse of modern civilization. Its creation in vast volumes seems inevitable and, once created, it must be disposed of safely. But is it inevitable? It is very true that, if there is no waste, there is no problem with waste. Our subject is waste management, but the best management approach is to manage affairs so that there is no waste to manage. That is of course an impossible dream, but the proper management of waste must begin with the question: can the amount of waste being produced be minimized, if not totally eliminated? The answer to this question is certainly in the affirmative. The production of waste *can* be drastically reduced in many instances. There are two main methods of approach to initial waste reduction: designing plants so that less waste is produced; and reusing or recycling such waste that is produced. We deal with the first of these approaches in this chapter, and turn to the subject of recycling, recovery and non-waste technology in the following chapters. Toxic waste presents its own special problems, and therefore continues to be the focus of attention from the experts at various international seminars and conferences. Typical of these was an international meeting at Vienna, where waste minimization was to the fore.[1] Whilst it was felt that much remains to be done with respect to waste treatment, by seeking out more efficient and cost-effective methods, yet a more direct approach was through waste minimization. The change in emphasis over the past 15 years was highlighted thus:

	Predominant Method
Fifteen years ago:	safe landfills
Five years ago:	safe treatment
Today:	waste minimization

Waste reduction is the way

Unfortunately, all the regulations relating to waste seem to be based on the premise that waste in unavoidable. There are no regulations prohibiting in any way the initial production of waste – only regulations stating what is to happen to it once it is there. They deal only with the safe disposal of waste once the problem has been created: a more fundamental issue is that attempts should be made and encouraged to reduce the volume of waste that has to be disposed of. There is no doubt that efforts to reduce and minimize waste can bring substantial dividends: not only is there a substantial financial inducement, but there are also other benefits that cannot be evaluated in direct monetary terms. For example, the 'image' of a company and its public relations will both improve. Furthermore, if, for instance, the volume of waste produced was halved, the problems associated with its disposal would be more than halved. Jane Bloom, a senior attorney with the National Resources Defense Council, has indeed been reported as saying:

> We're producing almost 1 ton/year of hazardous waste for every man, woman and child in America... the RCRA (Resource Conservation & Recovery Act) does not force generators to reduce at the source.... What is going to motivate generators – what has always motivated them – is a regulatory program and incentives.[2]

A research organisation specializing in environmental and natural resource issues, called INFORM, based in New York City, recently produced a comprehensive 535-page report on waste production in the chemical industry. This report concluded that the industry could do much more to reduce waste right at the source than it was doing. This is not only commonsense, it is also good business sense, since what you do not make you do not have to handle or process. If waste is eliminated at source, not only will there be direct cost savings, but there are benefits in terms of insurance liability. This is an aspect that is assuming ever greater importance, particularly since the Bhopal tragedy which highlighted the vital importance of the principle: what you don't have cannot hurt you![3] A recent Science Advisory Board report, 'Future Risk: Research Strategies for the 1990s' recommended a hierarchy of goals to implement pollution control programmes with waste prevention, not surprisingly, the paramount goal.[4] Only when residuals were unavoidable would the next step in the hierarchy be applied:

recycling and reuse. Finally, of course, where prevention and recycling were impossible, residuals would be destroyed through incineration. This concept is simple enough, but very difficult to implement. It is, however, the only route if one's liability is to be reduced to a minimum. Unfortunately, the issue of liability is completely indeterminate. Companies involved in the production and handling of hazardous waste, in particular, never know what may happen next. However much is done to ensure compliance with all the relevant rules and regulations there will always be hidden risks, bringing the lurking danger of an unknown liability in the future. As if this was not enough, there is the further danger that some regulations relating to toxic and hazardous materials have been made retroactive. This means that a liability is then created in respect of waste which was disposed of earlier in strict accordance with the regulations then in force. This is precisely what happened in relation to the notorious Love Canal. The company disposed of the waste with far better provisions for its safety than the current regulations required: in the 1940s it met the legal requirements of the 1970s. Yet it had to pay very substantial amounts in compensation. So what is the answer? The only completely effective answer is not to produce the waste in the first place! For this it is necessary to go back to the drawing board and review the plant and its processes to see to what extent the production of waste can be eliminated or reduced.

It seems that the chemical companies do indeed realize that this is a vital issue. According to one company spokesman, 'minimization is the name of the game'. There is no doubt that waste reduction is the way to go, and many major multinationals in the chemical industry are looking into the matter very closely, 'product by product and process by process'. All in all, waste minimization is the obvious and preferred choice, especially when waste is defined as 'valuable chemicals waiting to find a home'. There are many examples of this to be found in the chemical industry, such as di-basic esters, a byproduct of nylon manufacture, now sold as substitute solvents.[5] Whilst the chemical industry can do much through process changes and seeking outlets for their waste products, the problem will never entirely disappear, although its impact can be greatly reduced – indeed, *is* being greatly reduced.

The phosphorus plant built by Albright & Wilson in Newfoundland has become a classic illustration of short-sighted mismanagement. The plant is due to be closed in 1989, after a series of spillages that polluted local fishing grounds. First built to secure the cost advantage of cheap electricity, this advantage has been more than annulled by the distance of the plant from the company's home base and the hostile climate. However, the final closure of the plant illustrates the fact that chemical companies are sensitive to environmental issues. Indeed, some companies are making great strides with waste minimization. Since 1983 DuPont, a major multinational, has required each of its production sites to submit an annual plan to minimize waste, with targets for specific wastes, on the basis of what has been called the '5R' principle: reduce, reuse, recycle, recover and research. Whilst a great challenge faces such companies, it seems they have no option. Indeed, it has been asserted that the typical chemical plant at the turn of the century will have:

- A clean process, with recycling or reclamation of waste.
- No hazardous waste leaving the plant.
- Accidental and routine emissions considered a relic of the past.
- Community fully informed, with complete confidence in the plant management.

Figures speak louder than words

It seems that, as the result of the intensive efforts that have been made over the past decade in this direction, the volume of waste – particularly of hazardous waste – has been falling sharply. Less than half the volume of solid waste is being produced today than was the case at the beginning of the 1980s. For instance, the Chemical Manufacturers Association (CMA) in the United States carried out a survey of the solid waste management practices in relation to over 500 chemical plants across the country, which included 33 of the 50 largest producers there, operating in 1986. Data was provided for 221 of these plants over a five-year period, with the result that the trend could be clearly seen. This data is consolidated in Figure 10.1. It seems that of the total quantity of solid waste produced, 98 per cent was treated to clean water standards and discharged, leaving only 2 per cent hazardous solid waste, of which 70 per cent was recycled.[6]

The primary conclusions one can draw from the figures derived from the study is not only that the volume of solid waste has decreased substantially, but that it is less hazardous than it used to be. Further, this reduction has been achieved despite a steady

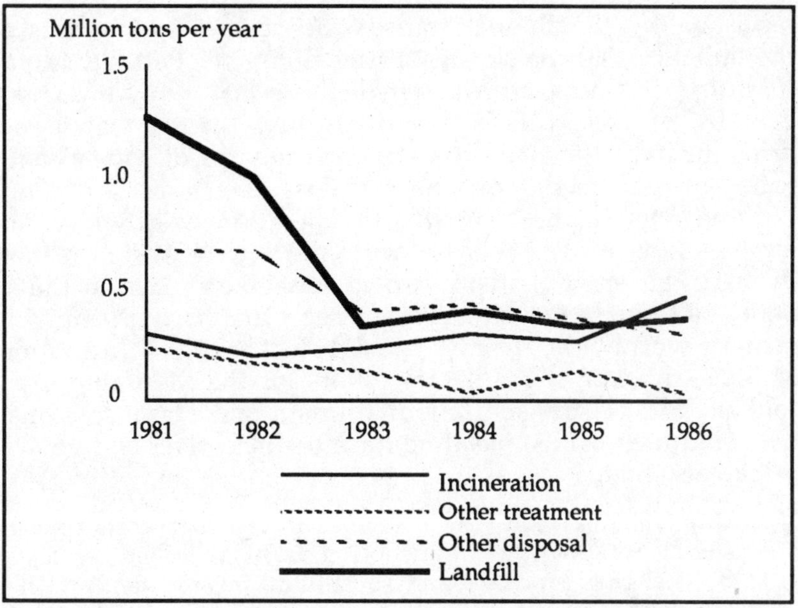

Figure 10.1 **Solid waste trends. These graphs indicate the trend in terms of volume and method of disposal for solid waste arising from 221 chemical plants in the United States that were the subject of a detailed survey. (Courtesy of the Chemical Manufacturers Association, Washington, DC)**

increase in industrial activity, said to be some 11 per cent over the period studied. It seems that not only are chemical companies producing less waste, but they are finding better ways of disposing of the waste they do produce. The data presented in Figure 10.1 shows a trend away from landfill for hazardous waste towards intensive treatment or incineration.

It seems that both the economics of the situation and stricter laws are driving chemical companies towards waste reduction. They face a shrinking capacity to take their waste, which is causing disposal costs to rise dramatically. They also have to face increasing public concern about the health risks associated with hazardous waste in certain cases, such as when it is disposed of by landfill. These are all compulsive factors pointing in the same direction: waste generation must be reduced. The CMA set up a waste minimization policy in 1987, designed to support their members and encourage activity in this direction. Such a policy should also serve to improve the public image of the chemical industry, badly tarnished by the Bhopal tragedy. So severe a blow was this to the industry that the slogan: 'Better things for better living *through chemistry*' has now been bereft of those last two words: can they ever be put back? It is our thesis that the very accidents that bring the industry such a disastrous image and provide a continuing feast for the media are nevertheless the basis for continuing improvement in the standards of operation worldwide. This certainly seems to be the case with respect to waste creation and disposal: substantial improvement has taken place in the United States, although we have yet to see what is happening in this context worldwide. Testifying on behalf of the CMA, Morton L Mullins, director of regulatory affairs for the Monsanto Company, told the House Transportation, Tourism and Hazardous materials Subcommittee, whilst pleading for more incentives and government initiatives, that:

> ...waste minimization equals savings in the high costs of managing wastes ... we save on production costs by recycling waste, as well as by changing processes and substituting raw materials ... it is also in our self-interest to generate less waste since this reduces potential legal liabilities associated with waste handling and disposal ... plants in our most recent survey recycled almost 70% of all of their solid hazardous wastes.[7]

But there is still scope for improvement

The quantity of waste being produced by industry seems to be falling, to some extent as a result of economic pressures, but also, fortunately, due to a voluntary movement motivated by the desire to improve and protect the environment. Nonetheless there is still tremendous scope for further action, and we feel that industry has, as yet, merely dealt with the 'tip of the iceberg'. Intensive efforts over a period of time might well reduce the volume of waste produced by industry itself to but one-tenth of the amount being produced at present. We say 'itself' because industry provides the public with many products, particularly packaging and wrappings, that eventually find their way into household rubbish. But so far as industry itself is concerned, recycling and the re-use of waste can well play a vital role, as we shall see when we take this subject up in our next chapter. Our modern society has for far too long accepted as inevitable what can in fact be drastically changed if the appropiate effort and resolution is brought to the subject. Perhaps the United States is the worst culprit in this respect, since it is said to produce twice as much waste per head as Japan and Europe.[8]

To facilitate re-use and recycling it is essential that waste be sorted prior to disposal, and the best place for this is at source. For instance, the domestic householder should be encouraged to sort waste so as to segregate, for instance, paper, glass and aluminium before it is discarded. Unfortunately no one yet seems to have devised a really effective scheme to encourage householders to do this. For instance, glass bottle banks are a common sight in many European countries, but their success depends upon the goodwill of the householder. There is no direct financial encouragement, except by providing industry with some financial inducement to minimize waste. The North Carolina state government set up a technical assistance programme with the slogan 'Pollution Prevention Pays'. They expressed a willingness to share the cost (up to US$5000) for the preparation of a pollution prevention report and were also prepared to finance pollution prevention projects, providing funds at a very low rate of interest. Indeed, attacking the problem at source in this way has proved to be both cheaper and more effective.[9]

Waste is generated at the drawing board

Prevention is always better than cure, so that when plants are being built, waste *creation* – or better, waste *elimination* – should be considered *before* waste treatment and disposal. This may well mean seeking an alternative process, an approach that may be time-consuming, but ultimately very rewarding. The problem must be attacked at its root and this could lead to its disappearing: it can be done. By way of example, despite the intensive efforts of US companies to dispose of the mounting piles of phosphogypsum, a byproduct of phosphatic fertilizer manufacture, in an acceptable manner, they continue to grow at a rate predicted to result in a billion-tonne inventory by the year 2000. It seems that, in the state of Florida alone, more than 350 million tonnes of phosphogypsum is stockpiled, while in Louisiana a task force of industrial, government and environmental groups is wrestling with the problem of the disposal of some 12 million tonnes. Now an engineering contractor claims to have developed a viable process to produce sulphur from phosphogypsum waste, converting the gypsum into elemental sulphur and calcium carbonate. The sulphur can then be used for the manufacture of sulphuric acid, while there are a number of options for the profitable disposal of calcium carbonate, one of which is its utilization as cement clinker.[10]

We can take another interesting example from the electroplating industry. Electroplating gives rise to very dangerous liquid wastes, but the Pioneer Metal Finishing Corporation of Franklin, New Jersey, has developed a 'closed loop' process that gives rise to a far less dangerous effluent. Thus they have an improvement which is not only technically feasible but economically viable. There are savings in the quantity of water used (since it is re-used), pollution control equipment is no longer required, the cost of waste disposal has been eliminated and there are lower maintenance costs. What is more, it is ecologically sound: there is a safer environment for both the employees and the local community.

There is no doubt that the fate of all such plants is 'sealed' at the drawing board stage. The process design determines how much and what kind of waste is produced at each process step, and it is here that elimination or minimization should be studied. This may be difficult, but the possibilities should always be reviewed. Minimization is usually possible and should be the objective. What is true for plant safety is also true for waste. In our book *Safety in the Chemical Industry* we have a chapter headed 'Safety starts at the drawing board'. We say there:

> The various aspects of safety in design should be taught and be an integral part of all engineering courses, at every level, with emphasis on the professional responsibility of the individual engineer.[11]

Substitute 'waste prevention' for 'safety' and the advice is pertinent to our present subject. Engineers-to-be should be brought to realize that they, personally, are responsible for the outcome of their work and its proper functioning. However incidental and minor their work may appear to be, they may well have the lives of people, perhaps many people, in their hands. Waste elimination has an importance in this context equal to plant safety, especially when the waste happens to be hazardous or toxic.

Regulatory reforms should stress the importance of the voluntary elimination of hazardous wastes 'upstream', where they are generated, rather than enforcing expensive controls 'downstream', as seems to be current design policy.[12] Almost everywhere, it appears, the present environmental regulations provide a 'no-win' situation, as more and more money is spent on the impossible dream of providing zero or near-zero risk. A sound approach to waste reduction demands a fundamental shift away from control, as at present, to the far more desirable goal of prevention. There have in fact been some notable achievements in this regard. Typical of what is possible was the process change adopted by Hewlitt-Packard. In the manufacture of their circuit boards they switched from a silk-screening process using organic solvents to a dry film process. Operating costs doubled, but the use of toxic chemicals was completely eliminated. What is more, the dry-film process enabled the manufacture of more sophisticated circuit boards. A very similar approach was used by Cleo Wrap of Memphis, Tennessee, the world's largest producer of gift wrapping-paper, who completed a six-year conversion project in 1986. They have replaced the organic solvent-based printing inks they were using by water-based inks, thus virtually eliminating the generation of hazardous waste in its plants. All the underground solvent storage tanks have been dispensed with, and a lower fire insurance premium has been obtained, due to the reduction in fire hazard now that there is no longer any storage or handling of flammable solvents.

Reviewing the possibilities

Examples are all very well, but each case is different. How then does one approach the subject? A number of valuable suggestions for waste reduction in relation to a wide range of industrial activity have been gathered together in one place in a paper by two members of the US Congress Office of Technology Assessment (OTA).[13] Its main conclusions were as follows:

1 Thanks to tradition, habit and the government emphasis on pollution control, industry has concentrated on the treatment of waste *after* it has been generated, rather than trying to *reduce* it.
2 Unless a comprehensive industry–government and mass media programme is drawn up and implemented, waste will not be reduced: pollutants will merely be moved around from the land to the sea or the air. Then no one will benefit.
3 There are meaningful ways of waste reduction. A waste by-product can be reintroduced as a raw material; production processes and operations can be changed; raw materials can be substituted; products can be reformulated; inventories of dangerous materials can be minimized.
4 Given the way in which companies are currently organized to deal with environmental problems, and the manner in which economic analysis is applied when evaluating alternatives, there is an in-built bias against waste reduction.

These are very valid points: those of our readers who are directly concerned in these matters should review them to see in what way their own actions are being inhibited.

The subject of waste reduction was apparently sufficiently important to prompt the US Congress to declare in 1984 that 'the generation of hazardous wastes is to be reduced or eliminated as expeditiously as possible'. But despite this, the subject does not seem to have been taken too seriously, even in the United States. Two years after the above declaration, of about US$16,000 million spent yearly by various US government agencies on environmental protection, only US$4 million (or 0.025 per cent) was devoted to waste reduction. Yet waste reduction could have a tremendous impact on the environment, if pursued with vigour.

The role of waste minimization

What is the difference between waste minimization and waste reduction? To have waste reduction as an objective is to seek to reduce the volume of waste produced. A realistic and serious waste reduction programme must start with a step-by-step waste audit and the compilation of the waste reduction possibilities. These are ranked in order that the optimum option may be selected. An extremely useful checklist has been provided in this context by Fromm.[14] The success of such a programme can be evaluated in terms of quantity of waste generated per unit of production, and the audit team must have a good understanding of the process and full cooperation from the plant operators.

The phrase 'waste minimization', however, as it appears in the 1984 amendments to the 1976 RCRA (Resource Conservation and Recovery Act) is used in a much broader sense than waste reduction, but *includes* waste reduction. It is meant to be all-inclusive, and so covers not only waste reduction but recycling and the treatment of wastes after they are generated. Waste minimization thus combines two basic concepts: prevention and control. Its ultimate goal is to have no land disposal of hazardous waste in recognition of the fact that land disposal, at present by far the most common way for the disposal of hazardous waste, does not completely protect human health or the environment and always poses hidden dangers. In the United States, this realization has recently led to a series of bans on the land disposal of certain classes of chemicals, and other industrialized countries will doubtless soon follow suit. Three waste minimization regulations, effective since July 1985, place the responsibility for the safe disposal of waste directly on the originator. This requires self-policing, a most effective technique. The three regulations, stated briefly, say:

1 When hazardous waste is shipped off the site of its generation, the producer must certify on the documents accompanying the shipment that they have a waste minimization programme in place.
2 Corporations who manage the waste of others or who generate and manage it themselves at the place of generation, must certify about their waste minimization programme being in place. This is to be placed in their permit file maintained at the site.
3 RCRA biennial reporting requirements apply to all the hazardous waste generators, and they are also subject to special waste minimization requirements. They are required to spell out in detail the waste minimization efforts undertaken by them during the year, and also report on any changes in respect of the volume and toxicity of the wastes during the period of reporting.

Thus management is made responsible for its actions. This concept has led to at least one paper on the subject with the unusual and provocative title: 'Exit the safety inspector?'[15] While the sole arbiter is legislation, with compliance assessed by an inspector, the company can feel absolved from responsibility: it has satisfied the legal requirements. Indeed, this is normally the defence made when things go wrong. Once, however, that the company is aware that a mere meeting of the regulations is no defence if disaster strikes, as is the case under the above regulations, it is far more likely to ensure that its provisions with respect to the disposal of waste are sound and safe.

Let us conclude, even at the risk of repetition, by pointing out once again that the problem of waste is best tackled at its roots. Waste that is not generated cannot create any problems, making non-generation the cheapest waste handling measure.[16] In the United States, under present RCRA rules, an informal target of 50 per cent reduction in waste generation as compared with that being achieved in 1986 seems to be very possible. A higher target than this may well be set up on the renewal of the RCRA. It has to be remembered that waste reduction programmes are highly specific and each company has to make its own assessment of the possibilities and then work towards their achievement. The employees should be actively engaged in seeking and promoting waste minimization: some of the slogans used in this context demonstrate the way in which such cooperation can be encouraged. For instance, Dow Chemical instituted a WRAP programme – 'waste reduction always pays' – while Chevron had their SMART programme – 'save money and reduce toxics'.[17] There are clear indications that waste minimization is going to be the dominant theme in waste management for the future. Further it is clear that the serious technical and financial challenges ahead in the waste management field can only be met by a strong commitment to research, educational programmes and intensive waste prevention techniques.[18]

References

1 Hunter, D., 'Vienna conclave focuses on toxic waste disposal', *Chemical Engineering*, **27**, April 1987, p. 17+ (2 pp.).
2 Rich, L.A., 'Hazardous waste management – new rules are changing the game', *Chemical Week*, 20 August 1986, pp. 26+ (24 pp.).
3 Kharbanda, O.P. and Stallworthy, E.A., 'The lesson of Bhopal', *Safety in the Chemical Industry*, Heinemann, 1988, ch. 9.
4 Alm A.L., 'Waste reduction', *Environ. Sci. Technol*, **23**, No. 3, March 1989, p. 271 (published by the American Chemical Society, New York).
5 Tilton, H., 'Waste reduction is the real priority', *Chemical Marketing Report*, 23 November 1987, p. 31 (part of *Waste Management '87*, a special 16-page supplement.)
6 'Waste management trends', *Chemecology* (cover feature), Chemical Manufacturer's Association, Washington, DC, June 1988, pp. 2–3.
7 Ibid.
8 'Rubbish – the burning question', *Economist*, **307**, 28 May 1988, pp. 39+ (3 pp.).
9 Keoningsberger, M.D., 'Preventing pollution at source', *Chemical Engineering Progress*, **82**, May 1986, pp. 7–9.
10 'Sulfur from phosphogypsum waste', M.W. Kellogg Company News Release no. 10, 1987, 7 April 1987.
11 Kharbanda and Stallworthy, op. cit.
12 Hirschhorn, J.S. and Oldenburg, K.U., 'Prevent pollution … upstream', *Chemtech*, **18**, May 1988, pp. 274–6.
13 Oldenburg, K.U. and Hirschhorn, J.S., 'Waste reduction – a new strategy to avoid pollution', *Environment*, **29**, March 1987, pp. 16+ (12 pages).
14 Fromm, C.H. *et al*, 'Succeeding at waste minimization', *Chemical Engineering*, **94**, 14 September 1987, pp. 91–4.

15 Baker-Conseil, J., 'Exit the safety inspector?', *Process Engineering*, April 1986, p. 29.
16 Basta, N. *et al*, 'Waste ... waste ... waste ... an ounce of prevention', *Chemical Engineering*, **95**, 15 August 1988, pp. 34–7.
17 Winton, J.M. and Riach, L.A., 'Hazardous waste management – putting solutions into place', *Chemical Week*, 24 August 1988, pp. 26+ (15 pp.).
18 Porter, J.M., 'Hazardous waste cleanup – A critical review', *Chemical Engineering Progress*, Vol. 85, April 1989, pp. 16–25.

Chapter 11

TOWARDS A NON-WASTE TECHNOLOGY

Non-waste technology is exactly that: no waste at all is generated. It is an issue of worldwide importance and has been regarded as significant enough to merit being defined by the United Nations Economic Commission for Europe thus:

> ...the practical application of knowledge, methods and means, so as – within the needs of man – to provide the most rational use of natural resources and energy, and to protect the environment.[1]

In the following chapters we shall be looking at the process of recycling and recovery. Whilst this is an effective way of treating waste, ideally there should be no waste at all. These concepts are demonstrated in Figure 11.1. Alternative (1) illustrates the normal approach in manufacture of a product, whilst alternative (2) illustrates a better approach, with the reclamation of at least part of the waste products. Alternative (3) illustrates the complete re-use of all the waste, obviously the best approach of all. Whilst that may be a dream, non-waste technology represents a positive, active and aggressive approach towards this most desirable end. The underlying objective is to eliminate, or at least minimize, waste at every stage in process development – design, construction, and operation – in relation to all industrial processes. Of these, design is the most importamt stage and is obviously the cheapest approach to the problem.

An economic and social necessity

The rapid industrialization that has taken place and continues worldwide has brought very serious environmental problems.

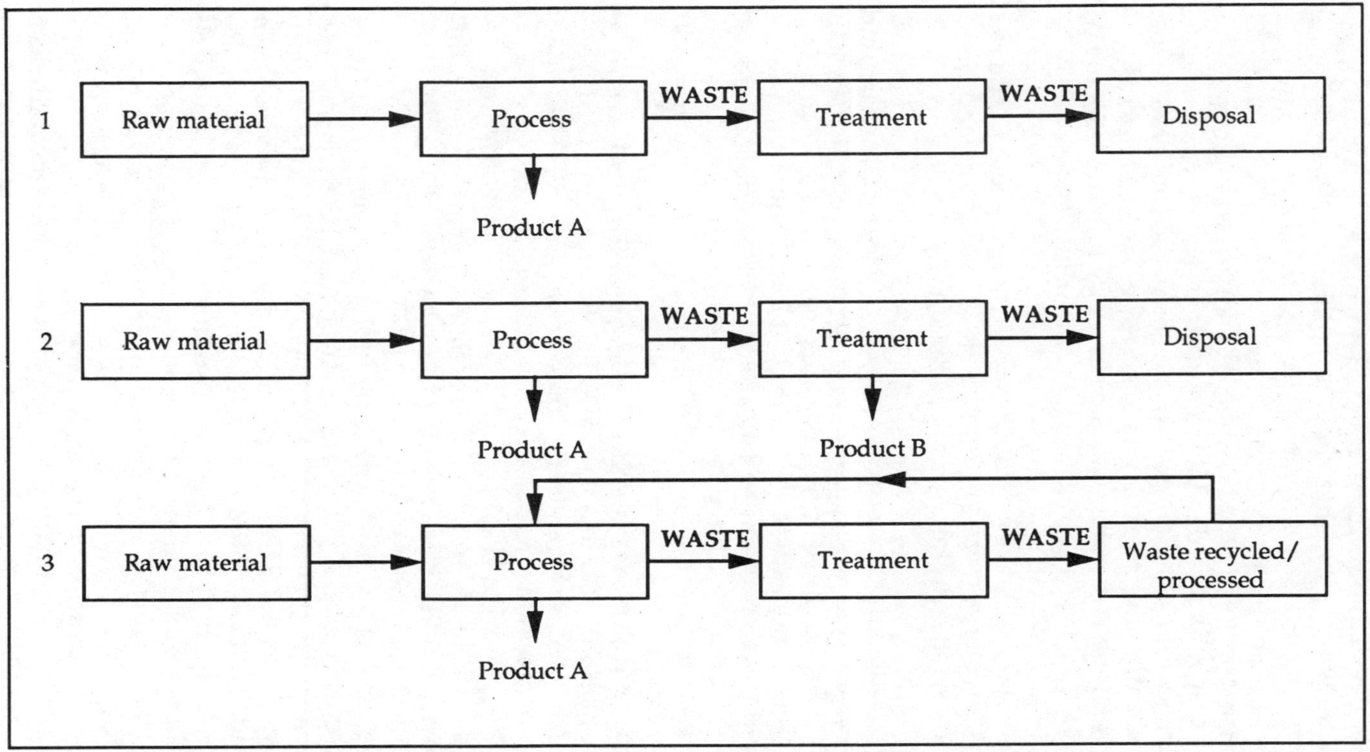

Figure 11.1 **The management of waste. These diagrams illustrate the difference between (1) normal treatment, (2) reclamation and (3) use and reuse.**

Advances in technology have resulted in the effective treatment of hazardous wastes, whether they be liquids, solids, or gases, so that the end result is a safe, but usually unusable product. To achieve this, costs have therefore been incurred which bring no benefit. Furthermore, this waste has to be disposed of and an additional cost is incurred which also brings no return. Recycling is an improvement on this, in that a usable product results, but it is still usually uneconomic. As we survey present trends, we see the development of a non-waste technology to be an imperative. The three key factors leading to this conclusion are:

1 Raw material resources are finite and are rapidly being depleted.
2 There is an energy crisis in terms both of availability and price.
3 There is serious environmental pollution.

These are all the direct consequences of a wasteful society, and the only real answer is to eliminate, or at least, minimize waste. In other words, we must move towards a non-waste society, for which we need non-waste technology. How should non-waste technology impact on the three key factors just mentioned above?[2]

- *Materials*: products should be designed for long life and conservation.
- *Energy*: every means should be adopted to avoid loss (e.g. by insulation) and to increase efficiency.
- *Pollution*: use non-polluting processes, and recycle where waste is unavoidable.

Much of the basic information and related technology for this already exists, but its existence needs to be brought to the fore and introduced into the curricula of both engineers and scientists. So basic is this concept to the eventual fully effective use of existing resources that existing curricula should be reviewed and revised to include this aspect. A non-waste technology already seems to be an economic necessity and it may soon become both a social and political compulsion. It can even be economically viable, especially on the basis of 'real costing' – an approach where *all* the costs are taken into account. This is particularly so because of the distorted way in which resources utilization has developed over the years. The developed world, whilst it only has perhaps one-quarter of the total population of the world and occupies less than one-quarter of the earth's surface, nevertheless is using more than three-quarters of the world's raw materials and energy resources. Worse still, much of this is being used unnecessarily: it is being wasted. To continue along this course can only lead to disaster. Non-waste technology offers a practical and viable solution to this growing dilemma, but it is a policy that needs to be adopted immediately if disaster is to be avoided. Will those in a position to act listen, before it is too late to take corrective action?

Non-waste technology is sometimes referred to as 'clean technology' (CT), since it enables production without pollution. But all too often this requires substantial modification to production processes. When deciding to invest in CT a manager must consider not only the environmental benefits, but also the economic and social factors associated with the innovative techniques it is proposed to use. The OECD has issued a report (see OECD Observer No. 148, October/November 1987) that studies the difficulties involved in making these techniques more a part of industry's standard equipment and the measures that have been taken by some countries to encourage their use.

The United Nations takes the initiative

The problem posed by unnecessary waste is basic, it has international connotations and is faced by everyone everywhere. As no place or people can escape its consequences, international cooperation should be an obvious course. In that event, the United Nations organization is an obvious medium. Recognizing this, The United Nations Economic Commission for Europe (UNECE) took the initiative and arranged the first international seminar on this subject in Paris at the end of 1976. Its theme was non-waste technology and production. A total of 75 papers from 34 member countries were presented at this seminar, and have been published in a single volume.[3] The papers presented fall under five main headings:

1 Concepts and principles of non-waste technology
2 State of non-waste technology: national and industrial experience, with case studies
3 Cost–benefit aspects of non-waste technology
4 Ways and means of implementing non-waste technology
5 Methodological and strategic aspects of non-waste technology

Those of our readers who wish to pursue this vital subject in detail should refer to the proceedings of this particular seminar, which contain a wealth of information on this new but fast developing field of endeavour. The main conclusions and recommendations that resulted from this seminar can be briefly summarized as follows:

- We must aim at both minimizing and eliminating waste.
- Every time the basic question must be asked: Do we need this product?
- A complete databank on non-waste technology should be developed, from the published literature.
- Human judgement and qualitative considerations must override the quantitive considerations.
- Non-waste technology conserves resources, apart from reducing pollution. Normal waste management only serves the latter purpose.
- Non-waste technology should therefore be considered as a positive philosophy and implemented in stages, with a long-term strategy in mind.

This seminar has made a major contribution to this new, vital, emerging discipline and, at the same time, has made immediately available both basic data and first hand operating experience from several industries in various European countries. We shall review some of this experience below.

The role of design education

We have already seen that the concept of non-waste technology, if it is to be effective, must be introduced into and integrated in the educational curriculum. Only in this way will the engineers and scientists involved in design be aware of the problems and so incorporate the appropriate solutions at the design stage. One of the papers presented at the seminar described above gives excellent coverage of this aspect of the subject.[4] Non-waste technology provides a long-term solution to the three problems, with which the world is faced, as defined above, when we were showing that non-waste technology is both an economic and social necessity. There can be no better approach than to review the situation and the possibilities at the design stage. The products, the devices and the machines of the future can and should be so designed as to incorporate non-waste technology. This places the engineer/designer in a key role, and to perform this function correctly requires training in the concept and philisophy of non-waste technology. However, this field is so vast that it is impossible to cover all the relevant aspects in a design curriculum, since consideration has to be given not only to the sound, economic design of the item itself, but also to its use and the consequences to others of that use. In view of this, the paper proposes an approach described as 'methodical design'. The objective is to build the philosophy of non-waste technology into the design activity, rather than seeking to cover all possible aspects of the subject. It seems that there is a great deal of very relevant knowledge on this already available: it only needs to be applied.

Methodical design, as its name denotes, is a systematic approach to design problems. It can stimulate creativity: indeed creative thinking, or better still, lateral thinking is a prerequisite if non-waste technology is to be incorporated at the design stage. The first phase of design has to take into account the environmental problems. In the second stage, the energy aspects should be reviewed and incorporated. In the third and final phase of design, the form and dimensions of the construction being developed are established. It is at this last, but crucial, phase that the materials to be used and the manufacturing methods to be employed are firmed up. To be able to minimize the volume of materials used, whilst at the same time minimizing the energy consumption and both the immediate and ultimate impact on the environment, the designer needs to take into account the following significant factors:

- *Materials*: the number of products, their lifetime and the amount of material required per product.
- *Energy*: the installed power required, the energy cost during the lifetime of the product, and the energy content of the product itself.
- *Environment*: The possibility of pollution, the total energy demand and the possibilities with respects to recycling.

The development and use of any new product follows what has been described as the 'bath tub' curve, which is characterized by three distinct stages:

1 *Infant mortality*. During this stage, the faults due to wrong design and poor production become evident and need to be rectified.

2 *Normal life.* the products, during this period, fail at a chance rate, which usually turns out to be fairly constant. This failure rate can be lowered by increasing the quality of the product, by more careful use through better instructions for use, and the incorporation of design features that prevent misuse.
3 *Wear-out period.* The failure rate increases due to wear, corrosion and fatigue. Better design and preventive maintenance can minimize this period, at the same time prolonging the life of the product.

Notice that from the beginning the primary battle is against faults and failure. The higher the failure rate the higher will be the waste of materials and the greater the problems created by that waste, apart from the unnecessary and escalating costs. It is very obvious that the failure rate is best controlled and minimized at the design stage. Thus non-waste technology will concentrate on the failure rate and its control, seeking to extend the life of the product. At the same time every effort will be made to minimize the amount of material in the product and the waste that will arise during its manufacture. For instance, if the number of parts in a product is reduced, not only will its reliability be increased, but it will be easier to maintain and have a longer life. Indeed, a better-quality product can be the result. Management should be committed to quality, but to achieve anything worthwhile, the subject has to be taken seriously.[5] In our view, the elimination of waste is an essential part of the 'quality' approach.

Waste can be reduced during manufacture by the selection of the right manufacturing process. Metal-forming wastes much less material than metal-cutting – 18 per cent as compared with 33 per cent. Similarly, drilling is a very wasteful process (48 per cent wastage) whereas extrusion only results in 12 per cent wastage. These are just two illustrations of the approach to design that should be adopted, and they apply largely to the light engineering industry. For heavy industry other considerations apply, but of course minimizing total waste becomes even more significant since it not only lessens the prime demand on materials, but also reduces handling and transport costs.

The use of energy also needs the most careful consideration. The total energy consumption includes not only the energy expended in the manufacture of the product, but the energy used in the manufacture of the raw materials, the energy consumed by the product during its lifetime of service, and the energy consumed in its final disposal. Each of these separate aspects should be reviewed carefully in order to ensure that the *total* energy used is kept at a minimum. Unfortunately, the designer may well be handicapped by shortage of data: there is very little systematic information available on the comparative energy consumption of various manufacturing processes. Manifestly, processes which involve the heating of materials, such as casting, hot milling, forging, heat treatment and paint-drying consume the most energy, but much more needs to be known than that.

Whilst it is obvious that the amount of materials used and the amount of energy expended at all the various stages in product development and use discussed above have a direct impact on the environment and the degree of pollution, this fact needs to be recognized and that impact assessed at the design stage: no later. Only then can corrective action be taken which will be really effective. When considering potential pollution, all the relevant aspects should be reviewed; not only direct pollution, but the impact of waste, noise, and heat disposal. What are these doing both to people and their environment? The choice of material can have an important bearing on its recyclability, whilst the pollution aspects of wasted energy are not always realized. Wasted energy is not only a waste of materials, but the release of that energy into the atmosphere contributes to pollution through the damage that can result. Thus there is not only a twofold loss, but unnecessary additional costs are also incurred.

A non-waste value system

Waste is enormously wasteful. It leads not only to pollution today, but to material shortages tomorrow. The age-old saying 'waste not, want not' is as valid in the affluent 'effluent society' of today as it ever was. The very existence of society as we now know it is threatened by the continuing waste of precious, scarce resources. If the present industrialized society is to survive, it must become a non-waste society with a non-waste economy using non-waste technology, and above all a non-waste value system. This is the message of a paper by M.G. Royston presented at the UNECE international seminar mentioned earlier.[6] This particular paper also provides us with another useful definition of non-waste technology. It is a technology 'based on the conceptualisation of the total system of raw material supply–production–consumption–disposal and recycling, viewed in an integrated and a systematic fashion so that no waste occurs'. But we must be careful: it is just not possible to have a world without waste. A non-waste technology will still only minimize waste; it will not completely eliminate it. Royston writes of ecoproductivity – a world where productivity is maximized, so that there is no waste. If waste could be prohibited by law, so that all sectors of the economy produced only those products that were basic to our needs – food, drink and shelter adequate to maintain us in health and well-being – then that would be the ideal of ecoproductivity. But this is an impossible ideal at this present time, although a most desirable objective to strive for.

Non-waste technology, properly applied, actually results in lower costs than the techniques in use at the moment, but that is only when the situation is reviewed comprehensively from the very beginning to the end of the processes involved. It must be so, since it involves no double handling of products, as occurs when there is waste. The main obstacle to the general adoption of non-waste technology seems to be the lack of the required imaginative effort. Business activities in relation to new products will demand diversification, and the necessary finance also has to be available. We already have some examples of the diversification that can be enforced by the pursuit of non-waste technology. They include:

- A concrete manufacturing company entering the leisure business;
- A clay company entering the construction industry;
- A whisky manufacturer entering the animal feed business;
- A municipality entering the raw materials and energy business;
- A port authority entering the oil business.

One common feature runs through all these illustrations: the goal is *not* a product plus waste, but a wider and complete range of products with no waste. Thus two major problems are solved simultaneously: valuable resources are conserved and pollution is eliminated. This is the way to go in principle, but there is still a long way to go in practice. It seems that, for an effective approach, radical changes are required in society – changes which are most unlikely to be accepted while sufficient raw materials are there to sustain the present approach. Take, for instance, a few major examples of the gross waste of energy that occurs with our way of life at the moment:

- A kilogram of animal protein takes ten times as much energy to produce as does a kilogram of vegetable protein.
- Air transport requires at least a hundred times as much energy as water transport per passenger or tonne kilometre.
- Per passenger kilometre a car requires 30 times the energy of a bicycle.

Consider the implications of a policy designed to reduce the consumption of energy in these areas. Nevertheless, non-waste technology demands that such alternatives should be seriously considered. The concept of 'no-waste' demands that only products that are useful to mankind should be made: this is a concept which then has to transcend all geographical and national boundaries. The avoidance of waste worldwide is indeed the only sound route to worldwide prosperity. Without it, some are always going to be prosperous to the disadvantage of others. Thrift, and the related word 'thrive', are key words in this context, both having a common root in the Old Norse word 'thrifta'. The dictionary defines thrift as 'saving ways, sparing expenditure', whilst to thrive is to 'prosper, grow vigorously'. We will thrive if we are thrifty, but this is a fact that is now largely disregarded.

A large Scottish distillery, producing whisky, was anything but thrifty. It was dumping its waste products into a nearby river, until it was threatened by the local authority to either clean up or close down. The company decided to clean up. The waste effluent was concentrated and dried and then sold as animal feed which proved to be highly prized and very nutritious. The cost of the plant to do this was recovered within six months and animal feed is now a very profitable sideline. But compulsion was needed: how much better it would have been for both the company and the environment, if non-waste technology had already been playing a part in their thinking.

Another illustration comes from the shipment of oil across the world. Empty oil tankers carry water as ballast, which then has to be disposed of when the tanks are to be filled with oil once again. At the port of Ashkelon, this oily ballast water and associated residues are pumped ashore and the oil recovered by passing it through highly efficient separators. This system allows a much faster turnround of the tankers, whilst the revenue from the recovered oil makes the installation self-financing. At the same time, of course, pollution from oily water is completely avoided. However such plants are few and far between: non-waste technology has still to make any real impact here. Indeed, its full impact would result in us ceasing to use fossil fuels, such as oil and coal, for the production of electrical energy. A nuclear power plant is much cleaner and far less polluting than a coal-burning power station. There is no ash, no carbon dioxide, no poisoning gases going into the atmosphere, and much less heat is wasted. The volume of waste is far less, although the safe disposal of the radioactive waste that is produced is the subject of much debate.

When considering the various sources of energy available to man, undoubtedly the best, in terms of non-waste technology, is solar energy – not in the generation of electrical energy by the use of photocells, but by the proper management of the world's forests and arable areas. Proper management, with energy being produced either by burning or chemical and biochemical processing, could produce enough to satisfy the entire energy needs of the world. If tropical forests were cultivated for fuel, this would form an ideal example of non-waste technology. Even the ash residues are a valuable fertilizer. The solid and gaseous products would be used as fuel, the oil fraction as a biochemical feedstock. The organic chemical products produced could include polymers, used for products designed in their turn for recovery and recycling. This is then a full integrated agriculture–fuel–industrial complex, and is a concept just waiting to be used. It requires no new or unknown technology, and is already being practised to some extent. However, it requires global exploitation.

Examples of the practical application of non-waste technology can be multiplied, but we think that enough has been said to demonstrate the practicality of the approach. At the same time, radical thinking is required for its widespread use – thinking so radical that we doubt whether much will be done until there is dire necessity. It seems that, in general, we are only prepared to learn the hard way, continually ignoring the lessons of yesterday.

References

1 UNECE Proceedings: *Non-waste Technology and Production*, Pergamon Press, 1978.
2 Kharbanda, O.P., 'Need of the day – non-waste technology', *Chemical Industry News*, September 1982, pp. 390–1.
3 UNECE Proceedings: op cit.

4 Van den Kroonenberg, H.H., 'The role of design education in non-waste technology', UNECE Proceedings: *Non-waste Technology and Production*, Pergamon Press, 1978, pp. 583–99.
5 Crosby, P.B., *Quality Without Tears*, McGraw-Hill, New York, 1984.
6 Royston, M.G., 'Eco-productivity a positive approach to non-waste technology', UNECE Proceedings: *Non-waste Technology and Production*, Pergamon Press, 1978, pp. 39–52.

Chapter 12

RECYCLING: THE PRINCIPLES

Waste is initially generated at the drawing board so it is there that the problem should first be tackled. Part of the approach must be to explore the possibility of recycling and re-using some of the byproducts which would otherwise be waste products. This does not necessarily have to occur within the plant; the 'waste' can be sent to others, although this may mean it requires some form of treatment prior to despatch. This can be a very profitable business: one company in this business, Allwaste Inc., actually declared: 'one man's garbage is another man's gold'. There seems to be no doubt, therefore, that the recycling of waste, whatever its form, is a very sound approach to a mounting problem. The collection and disposal of domestic waste has now grown into a perplexing challenge, with an increasing population and costs of handling and transport also growing fast, and ever more exacting regulations to be complied with. At the same time, the disposable-based once-through economy, of which the mounting volume of waste is a prime indicator, is serving to steadily exhaust valuable, non-renewable resources very quickly indeed.

Waste not, want not

The immediate and most direct answer to the problem of the growing mountain of waste is to recover and recycle to the maximum extent possible. The National Center for Resource Recovery (NCRR) was set up in the United States in 1970 to explore such practical and sensible alternatives to waste disposal, and this example has been followed elsewhere. Similar organizations have sprung up elsewhere with the same objective in view.

There is no doubt that many basic materials, such as metals, paper, and glass, can be separated out and recycled to the advantage of us all. At the same time the natural raw materials – mineral ores, timber and limestone respectively – are being conserved.[1] According to a NCRR estimate, typical domestic waste as generated in the United States consists of:

	%
Paper:	35
Food and garden waste:	31
Glass:	10
Metals:	10
Plastics:	4
Rubber, leather, textiles etc.:	10

We would expect that the volume of plastics has increased substantially, largely at the expense of the wastepaper and glass. Other industrialized countries can be expected to show a very similar pattern in relation to their domestic waste. The NCRR made one very interesting proposal given that government offices have a reputation for generating vast quantities of paper. It encouraged the Energy Department to experiment with the conversion of waste paper from the federal offices into fuel pellets to fire the Pentagon boilers!

The plant handling waste to prepare it for further processing is becoming highly sophisticated. There are massive machines which ingest the waste, classify it by size and weight, shred, sort magnetically, air blast for further separation, and then treat it with chemicals. Recycled materials are used for the manufacture of cans, bottles, paper and the like. Solid residual waste can be used to generate energy in the form of steam, gas or electricity – an aspect discussed earlier. The recycling of metals, in particular, can lead to enormous savings in energy as compared with the initial manufacture. For instance, the recycling of copper consumes a mere 13 per cent of the energy required for the manufacture of virgin copper – a saving of 87 per cent. With aluminium the saving is even greater – 94 per cent! In addition, both air and water pollution will be reduced by recycling, and less water will be used. The reduction in impact on the environment achieved by the recovery of aluminium, steel, paper and glass is presented in Table 12.1.

Table 12.1 Benefits achieved by recycling

	Reduction Achieved (%)			
	Aluminium	Steel	Paper	Glass
Energy use	90–97	47–74	23–74	4–32
Air pollution	95	85	74	20
Water pollution	97	76	35	–
Water use	–	40	58	50

The major users of such facilities are, inevitably, the municipalities, since they have responsibility for the disposal of domestic waste. Some of the basic principles that should be considered when municipalities are considering resource recovery are as follows:

1 Use proven technologies. This will avoid trouble, such as occurred at Baltimore, where dynamite had to be used to break up tonnes of shredded material that had been massed together by spontaneous combustion.
2 Very large quantities are required before the process becomes economic. For modest quantities, landfill remains a much cheaper option, although it is not the preferred course since the waste is just 'wasted', not used. It is also hidden, and not always safe.
3 Ensure that there are buyers for the various products from such plants. For instance, there is a general reluctance to use refuse-derived fuel since it is felt to be too variable in quality.

Because of its size, a resource recovery plant calls for a very substantial investment, and such projects are not free from risk. They are not therefore very attractive to the average municipality, even though they are the most logical users of such plants. It seems that, whilst resource recovery is strongly promoted, it is weakly supported at government level.

From trash to cash

Despite what we have just said, it seems that there *are* profits to be made from waste recycling. Indeed, one article on this subject carries the headline: 'A new generation of entrepreneurs is making good money in the "hippie" business of recycling'.[2] This headline is revealing in more ways than one. The use of the word 'hippie' (someone who behaves unconventionally) indicates the type of image waste-handling has in the public eye – that is, not really respectable. Whilst the re-use of waste has been promoted as good economic sense for the public, it can also be good business for those who engage in it. For instance, Jack Lupas set himself up in Atlanta collecting used computer paper (not waste paper, only used computer paper). He collects some 200 tonnes per month from about 150 firms scattered across Atlanta, buying it at around 7 cents/kg and then selling it to a broker at four times his cost price. The broker, in his turn, sells it with all his other waste paper at a good profit to a giant paper manufacturer, such as Kimberley Clark, who convert it into towels, napkins, paper handkerchiefs and the like. The ever-increasing cost of waste disposal and the fact that there *are* profits to be made from recycling have fuelled the current boom, but far more can be done.

Nevertheless, there is enough activity in the field of recycling to warrant a number of journals dealing with the subject, such as *Recycling Today, Resource Recycling, Resource Conservation* and *Conservation Recycling*. Paper recycling has been encouraged by promoting the motto, 'recycle your paper, save a tree', but this approach has an inherent weakness since it depends very heavily on being able to encourage the public to cooperate: there is no direct financial inducement. That was where Jack Lupas made his mark: he offered to pay for used computer paper and thus encouraged its separate collection. However, new solutions are continually being found to old problems. The disposal of old tyres has posed a serious problem for many years. Now, it seems, they are being ground up to form 'crumb rubber', which can then be used in rubber pavements. Apparently, in certain circumstances, the use of 'crumb rubber' results in a better, more durable, product than does virgin rubber. As we survey the scene it seems that it is all a matter of economics. Put very simply, recycling becomes attractive as the cost of waste disposal by other means rises. Whilst it is profitable, it will be pursued. It would be far better if it were used because it minimizes the damage that waste does to our health and the environment, but that will never happen. The profit motive will always prevail. Can the two be combined? A good environment could well be a good investment.

Because recovery and recycling techniques are both process and material–specific, we cannot generalize and so present a broad picture of what is happening in this industry and the way it is developing. Each possibility has to be studied and assessed on its own merits. Nevertheless, by way of illustration, let us mention a chemical, polyvinyl alcohol, used in the sizing of textile yarns. This chemical has to be scoured from the cloth before it is dyed, is washed away in the effluent from the dyehouse and so released into the environment. However, a hyperfiltration process has now been developed which recovers some 96 per cent of the polyvinyl alcohol from the effluent stream, so that it can be recycled. Polyvinyl alcohol is not biodegradable and it therefore pollutes the environment. That can now be prevented. The process appears to be in use, so we assume it to be showing a profit. The extent to which it is used is liable to depend on the profit inherent in the process, unless stricter regulations enforce its adoption on a wide scale. This is the continuing problem with all such recycling processes: even when a solution has been found, it is not necessarily adopted. In some cases, whilst the generator of the waste may not be able to re-use or recycle it, others can but he does not know of them. This has led to the setting up of waste 'clearing houses' which bring the two parties together to their mutual benefit. Because the nature of the waste and its chemical composition has to be disclosed in full detail so that the potential user can assess its possibilities, this could reveal the process being used. Thus, if necessary, in order to protect trade secrets, the identity of the supplier can be withheld from the user of the waste. Such clearing houses are often provided as a service to their members by neutral and non-profit making organizations, such as trade associations and chambers of commerce.

Recycling: the potential

There is every possibility that at least half of the domestic waste currently being generated could be recycled. The potential is there, but to achieve such a degree of recycling requires considerable planning and intensive effort. Paper, for example, will lose its value rapidly when mixed with organic food waste. Glass and metals are less vulnerable to degradation in this way, but they still need to be segregated. Organic waste, if destined for composting, must be free of inorganic substances toxic to plant life. In general, the nearer the recovery process is to the source of the waste, the less sorting and processing will be required before recycling is possible. The enormous savings in energy that can result from recycling metals has already been mentioned earlier.

Table 12.2 The use being made of recycling in some of the major industrial countries

	Percentage Recycled (in 1985)		
	Aluminium	Paper	Glass
United States	28	27	10
United Kingdom	23	29	12
Japan	32	48	–
The Netherlands	40	46	53
West Germany	34	40	39

The extent to which recycling is being used in several of the major industrialized countries is presented in Table 12.2. Local factors will account for some of the variations to be seen in this table. For instance, the very high recycle percentage for glass in The Netherlands is, we believe, accounted for by the fact that not only has the bottle as a container been largely retained there, but the price of the product includes a nominal amount for the glass bottle, which is recoverable when the bottle is returned to the shop. Most supermarkets there have a machine installed in the shop which accepts such bottles automatically and delivers a 'print-out' which can then be cashed on passing through the check-out.

In considering the impact of recycling, the most significant aspect is the reduction in the demand for the basic raw materials, and the reduction in pollution from the manufacturing process. For instance, the recycling of a tonne of aluminium, apart from the enormous saving in electrical energy, eliminates the need for four tonnes of bauxite and 700 kg of petroleum coke, whilst the emission of the air-polluting aluminium fluoride is reduced by some 35 kg. It seems that the recycling of aluminium is growing apace. In the last 10 years Austria has more than tripled, and Japan more than doubled, its recycling rate for aluminium. The recycling of glass is another area of rapid growth, apparently having increased more than 50 per cent over the past five years in a number of countries, including the United Kingdom and West Germany. Obviously the recycling rates for such materials as these could easily exceed 50 per cent, if only the appropriate measures were taken. Indeed the OECD has estimated that, in the countries for which it presents statistics, over 90 per cent of waste glass could be made available for recycling. The potential is there, but the will and the determination to exploit it seems to be lacking.

Paper is another case where the impact of recycling on the environment could be quite dramatic. Recycling paper not only spares millions of hectares of trees from felling, but also conserves energy and reduces water pollution, as can be seen from Table 12.1. The scale of the problem can be appreciated once it is realized that using recycled paper for the printing of just one Sunday edition of the *New York Times* saves some 75,000 trees. If only half of the paper used in the world today were recycled, that would not only meet a substantial part of the demand for new paper but also preserve nearly 8 million hectares of forestland – an area equivalent to about 6 per cent of all Europe's forests!

The recycling of plastics is now receiving substantial attention. It seems that there are a number of factors driving the plastics industry into recycling, despite the technical difficulties. Apparently there are many plastic products that can be modified to make them degradable and this is obviously a very desirable approach. There is a new potential for degradable plastics in the light of the growing impact of environmental factors and coming world-wide legislation which is likely to present obstacles to the future growth of the plastics industry.

Recycling has unfortunately been hampered for years by the premise that it should be financially viable. This is a mistake. Recycling should be seen as a cost-effective 'disposal' option, especially if it requires lower government subsidy than other alternatives, such as landfill or incineration. In fact, recycling deserves maximum subsidies, since everyone will gain. What really matters is the energy saving, the resource conservation and the cleaner environment that will result, but unfortunately these have no direct, immediately apparent commercial benefit: certainly not for the company involved in recycling. Another problem is the fact that an economically sustainable recycling programme requires a high degree of consumer participation, and this can be most difficult to ensure. Incineration is an approach to waste disposal that has also inhibited the full development of recycling. In general, not only can incineration pose environmental hazards, but wastes that are more valuable when recycled should be recovered rather than burnt. However, the general tendency, particularly with the municipalities, has been to overbuild incinerator capacity in order to cater for the once-through disposal mentality and culture that now prevails. The plants that have been built have to be fed if they are to be operated economically, and the effort to maintain plant loading curtails any incentive there might be to sort and recycle the waste. A better approach would be to maximize recycling capacity, leaving incineration as a secondary option, rather than the primary option.

Recycling on a large scale demands a new way of thinking for those involved in waste disposal. Unfortunately, its successful implementation requires the involvement of a number of separate and often widely dispersed participants and this can be very difficult to organize. Nevertheless it can be done, and the public *can* be motivated to cooperate. Demanding deposits on beverage containers, which are refundable if they are returned in a clean condition, has helped give impetus to their recycling. A number of industrialized countries, including the Netherlands, Scandinavia, the Soviet Union and some US states, use this technique to some degree. The technique also operates to a limited extent in Australia, Canada, Japan, West Germany and Switzerland. Aluminium can manufacturers in the United States have also been fairly successful in promoting the recycling of their cans. It is estimated that more than half are now being returned, but the public have to be encouraged by some form of reward, even if indirect – in this case, payments to charitable organizations.

Efforts in this direction are not so well organized in the developing countries. Informal early morning collections of specific waste can be seen in most areas, especially the major cities, and it provides a livelihood for perhaps a few million people, being highly labour-intensive. However, it is usually well organized, being based on a network of buyers with appointed neighbourhood agents, some specializing in one or more categories of waste material. In Bangkok, Thailand, collection crews spend their time to and from their normal place of work collecting and sorting paper, bottles, cans and plastics. Earnings from such activities can often be as much as from a normal day's work. The governments of Indonesia, the Philippines and some other countries are positively encouraging such activities, since it helps to conserve their precious foreign exchange, at no cost to the exchequer. The scale of such operations, although largely implemented with casual labour, can be significant: municipal waste tips in Mexico City are said to support some 10,000 people.

Towards a recycling society

For the best results, a recycling philosophy needs to be built into the overall integrated plan for urban development and its waste disposal. When 'added on' as an afterthought, it can prove expensive and hence is far less attractive. Recycling has now come of age: the plant and the equipment are there, but there is an urgent need for full-time recycling coordinators at both state and municipal level to ensure that full advantage is taken of the possibilities. The same thinking needs to be applied worldwide. Cities can encourage recycling by offering the appropriate franchises, but they have to stipulate that it starts at the curbside. A few cities in the United States have already adopted this approach, but it needs worldwide acceptance if it is to have the proper impact. Fiscal means, such as imposing a heavy fee on landfill or incineration, will also encourage recycling. As the volume recycled increases, the process will become more profitable due to economies of scale and the establishment of stable markets. Governments can encourage recycling by a variety of means, such as requiring that the companies who receive their contracts use a certain proportion of recycled paper. Once again, as the demand grows and volumes increase, costs will fall. The technical press is also making an ever-increasing contribution to developments in this field, publishing articles with challenging titles such as 'Moving recycling to the front line' and 'Can markets for recycling be mandated?'.[3,4]

A detailed study of the requirements of industry, with an assessment of those needs that could well be satisfied by recycled materials, could also stimulate recycling. The potential is there, but creative thinking is called for. Unfortunately, today's tax codes, pricing mechanisms and marketing practices tend to discriminate against using recycled products. Any assessment of cost completely ignores what is in fact a vital aspect – the product's afterlife. The problems of disposal and the possibilities for re-use are either ignored or said to be someone else's responsibility. The entire process of a product, from beginning to end, needs to be assessed from a cost point of view – and by 'end' we mean when it is discarded and has to be disposed of. If that involves cost to the community then it is, in fact, part of the cost of the product.

There is yet another aspect which discourages, or at least fails to encourage recycling. In most cases, raw materials, energy and other inputs to production are rarely costed at their full, that is their societal, cost. Depreciation makes a mockery of the costing process in that it encourages the wasteful consumption of virgin materials when a recycled product would not only serve the purpose just as well but would also minimize pollution and conserve valuable natural resources. Indeed, the very subsidies designed to encourage proper pollution control by reducing its cost work against recycling in that, if it was more expensive to dispose of waste, then recycling would appear more attractive in terms of cost. Even water is normally sold below cost and certainly far below its replacement value. All such factors encourage waste and inhibit recycling. So a really radical approach to the subject of cost is required, with products and packaging designed to use energy efficiently and minimize waste. Plastics are an outstanding example of the misdirection that has taken place in this area. Seemingly a boon, they have become a bane. Their recycling potential is far from being realized even though they have largely replaced materials such as paper and glass in packaging – materials which are readily recycled. These are the sort of factors that should be taken into account if recycling is to be effective. The next chapter shows how this works in practice.

References

1 Thoryn, M., 'Resource recovery means waste not, want not', *Nation's Business*, **67**, May 1979, pp. 95+ (3 pp.).
2 Schwartz, J., 'Turning trash into hard cash', *Newsweek*, 14 March 1988, pp. 38–9.
3 Gubbels, D., 'Moving recycling to the front line', *Biocycle*, **30**, No. 3, March 1989, pp. 61–3.
4 Krishner, D. and Ruston, J., 'Can markets for recycling be mandated?' *Biocycle*, **30**, No. 2, February 1989, pp. 64–7.

Chapter 13

RECYCLING: THE PRACTICE

Having examined the continuing development of the techniques involved in recycling waste, we now propose to see how this has worked out in practice in various countries of the world. The issue, it seems, is important enough to warrant widespread cooperation on an international scale. Typical of the activities in this area was the International Recycling Congress, the fifth of its kind, held in Berlin early in 1987. This conference was by far the largest to date in this growing field, with some 2,200 delegates from 44 countries, who participated in the technical meetings that were held. In addition, there was an exhibition of relevant plant and machinery, where some 70 major manufacturers and service companies displayed their wares.[1]

The venue of the Congress should not surprise us: West Germany is particularly concerned about the problems associated with the disposal of its waste, which are substantial. The country has a population of some 60 million, and it produces 340 million tonnes of waste per year – some 5.5 tonnes per person per year, or 15 kg per person per day. When we reduce such figures to a daily rate, the magnitude of the problem becomes, we feel, much more apparent. Continuing with this particular example, this annual volume of waste consists of:

	Tonnes
From:	
Industry	119
The construction sector	72
Mining	69
Clarification processes	36
Domestic waste	30
Miscellaneous	14

These figures exclude a further 200 million tonnes of agricultural waste which is normally ploughed into the soil as fertilizer.

There is, however, worldwide recognition of the magnitude of the problem and the need for action. For instance, the multinational company Monsanto have now (1989) set themselves the target of reducing toxic air emissions from their plants throughout the world by 90 per cent by 1992. But reducing air emissions can increase the emissions to water, land or other media. So the company has also opted for a general waste *elimination* programme. All the operating divisions are intensively studying methods for outright elimination and recycling. This will take longer to achieve but is far more desirable environmentally. Despite the initial cost, it is nevertheless believed that company earnings will ultimately benefit.

Recycling can make a very substantial contribution towards minimizing the volumes of waste that have to be disposed of, but it requires continuing, sustained effort. Industrial waste is comparatively easy to recycle, given proper planning and foresight, but some wastes, such as clarification waste, excavated soil and spent oil have presented a challenge. If the generator of the waste cannot recycle his waste himself, he needs to look for a buyer to take it from him. Waste exchange is a means of using waste from one industry that is valuable to another. Exchange can be difficult, but the West German Industry and Commerce Chamber has taken a lead in solving this particular problem by cooperating with its counterparts in Austria, France and The Netherlands to set up a Waste Exchange. The exchange produces huge catalogues containing information about all sorts of used materials. At buyer-seller meetings between the four member countries, transport arrangements are negotiated and prices fixed. The exchange has helped to recycle a wide range of materials, including used synthetics, paper, glass, lead, packaging materials, timber, flyash and chemical waste. There have been other attempts to get buyer and seller together: for example, a comprehensive encyclopedia has been published to serve this purpose.[2] Not only is the waste producer happy to get rid of the waste, but he may even get payment for it.

One man makes all the difference

Whilst everyone seems convinced that recycling is the way to solve the problems associated with waste disposal, it seems that this is largely 'lip service'. What is actually needed is positive action, but such action involves the community, and community cooperation can be most difficult to achieve. However, Barbara Goldoftas, who teaches at Harvard University, tells an inspiring story of successful recycling which revolves around just one concerned citizen, town councillor Greg Bohosiewicz.[3] The small US town of Wilton, in New Hampshire, was using landfill to dispose of its waste, and there were no manifest problems until the mid-1970s. Then:

> ...the state began pressuring Wilton to close the landfill because it sat on the banks of a river. The town faced a number of options, none of them easy: open a proper landfill, export the waste, incinerate it, or recycle it. Enter Bohosiewicz, then 'just an interested citizen'.[4]

Bohosiewicz knew little about the technicalities, but he wanted to keep the local taxes down. As an economist he realized that recycling could well be both the best and the cheapest alternative. Operating as a 'one-man committee', he sold the idea of a joint recycling centre to six towns around Wilton. The centre opened in 1979 and accepted a wide range of materials, such as aluminium foil, bottles, cans, plastic milk containers and food waste. By 1987, more than half (65 per cent) of the area's residents were active participants and their waste disposal was costing them some US$36 per ton, whereas a neighbouring town that did not recycle was spending US$120 per ton. Wilton's recycling scheme was a success because its residents were cooperating fully, separating a wide range of materials for recycling. Nearly half the waste handled is recycled, whilst 35 per cent is incinerated and 15 per cent, together with the incinerator ash, is disposed of by landfill. The residents are required to bring all their refuse to the recycling centre and those who do not (some 35 per cent) must pay to have their waste taken away for safe disposal. But the credit goes to Bohosiewicz who succeeded in getting the residents of Wilton and the other towns who joined the scheme to participate. The receiving centre is said to look like a summer camp, with a tidy, litter-free parking area, and signs specifying what goes where. What is more, there is hardly any unpleasant odour.

Whilst recycling has only recently emerged as an option in the United States, it is already a key process for waste disposal in Europe and Japan. Japan, it seems, has the world's most successful recycling programme, with some 65 per cent of municipal solid waste being recycled. Taking Machida City as an example, it is located some 60 km south of Tokyo and has a population of some 620,000. City officials go door-to-door with sanitation workers, at least once a year, explaining the purpose behind the separation of waste. Residents separate their waste into a range of stipulated categories, such as newspapers, glass bottles, aluminium and steel cans, combustibles and non-combustibles. The inhabitants receive allocations of tissue paper, napkins and toilet paper in exchange for their newspapers. Non-recyclable waste is burnt, supplying heat to a cultural centre, its greenhouse and swimming pool.[5] So it can be done, and small communities like Wilton and Machida have shown the way.

Progress in the United States

The US government is beginning to encourage waste reduction and recycling at the state level. An EPA task force has laid considerable emphasis on this and has set the following goals for 1992:[6]

	Present %	1992 %
Landfilled:	80	65
Recycled:	11	20
Incinerated:	9	15

To achieve these goals, states and localities must set up integrated waste management systems, mixing and matching the options appropriate to their needs. Landfill should be the last resort.

Whilst waste recycling is making steady progress in the United States, increasing the extent of recycling is hard work, requires legislative action (sometimes unpopular), and a high degree of political will, but it can be very profitable in the sense that the total cost of waste disposal is reduced. This was the prime motivation in the cases of Wilton and Machida. One of the magazines devoted to this subject, *Resource Recycling*, sees the steady growth of recycling as due to a gradual realisation that although recycling costs money, it is less costly than the collection and disposal of waste.

What must be realised, however, is that cost should not be the prime motivator. Recycling is the best solution in relation to the environment, and concern for the environment should have a powerful influence on the decisions that are reached. It also happens to be by far the best overall option for the national economy. At the moment, practically all of the recycling taking place in the United States is organized by the municipalities, spurred on by the ever-increasing cost of landfill. As a result, Wilton-type programmes are increasingly becoming a part of a growing national trend towards recycling. At the last count known to us (mid-1987) at least 13 states, mainly in the north-east and on the West Coast, have passed legislation promoting the recycling of household waste, or making it to some degree mandatory. Over 500 communities offer roadside collection of glass, metal, paper and some other materials. California has laid it down that bottle-return and redemption centres be set up near major supermarkets. Encouraged by such developments, the EPA expect, as indicated above, that by 1992 perhaps 20 per cent of America's waste will be recycled.[7] Similar movements in Japan and elsewhere have also become popular and need to be encouraged in every possible way.

To illustrate the way in which individual companies can contribute to the overall effort, McDonald, the well-known fast food chain, have initiated a pilot project to test the possibility of recycling polystyrene waste from 20 of their restaurants in New York City and Los Angeles. Mobil Chemical are now well along in the business of recycling plastics, with a separate group, the Solid Waste Management Solutions Group, whose task is to identify and exploit business opportunities offered by plastics and other materials.

It has to be realized, however, that there are limitations other than the enthusiasm of the householder to effective recycling. Much also depends upon the manufacturer. Cooperation from the manufacturer is very necessary if the products which can use recycled material are to be made. Fortunately, the manufacturers of aluminium, such as Reynolds and Alcoa, are most ardent advocates of recycling, and are very ready to receive reclaimed cans. It is estimated, for instance, that more than half of the billion billion beverage cans sold in the United States since 1981 have since been put in recirculation, with a recycled can coming back on to the supermarket shelf within some six weeks. However, paper, unlike aluminium, cannot be recycled indefinitely, since the fibres eventually break down. Nevertheless, despite this, of the 600 or so paper mills in the United States, nearly one-third are based entirely on waste paper, whilst another 300 use from 10 to 30 per cent waste paper in their pulp mix.

Recycling elsewhere

Countries such as West Germany, The Netherlands and Japan, who do not have an abundance of raw materials and are also limited for space, have used recycling for many years. Most of the recycling takes place with municipal waste, with more than 50 per cent of such waste being recycled. Waste, once deposited, is normally the legal property of the municipality, but exemptions to this rule have been made to facilitate recycling. For instance, a day is nominated for surplus or unwanted furniture and other household goods to be placed at the roadside, and anyone is at liberty to remove such items for their own use before the council cleaners come along. The main factor in the success of recycling programmes in such countries has been the education of the public, in order to win their cooperation. It is recognized that what may be waste and not wanted by one may well be riches for another. This is almost always the case, as we have found by personal experience. One of us picked up a discarded portable typewriter in the United States in this manner, took it home and had it serviced. It is still in excellent condition after five years further use!

In sharp contrast to the attitude that prevails in continental Europe, Britain is seen as the 'dustbin of Europe'. Not only is its attitude seen as 'lazy', with vast quantities of waste being dumped into the sea – very convenient with an island community – but Britain seems very willing to accept waste from abroad for processing. Indeed, Britain is perhaps the only country in the developed world to make big business out of importing waste.[8] As in other countries, waste in Britain is treated either by the waste creator or by contractors, which have their own association – the National Association of Waste Disposal Contractors. There are perhaps 10,000 such companies, mostly small-scale (one man and a van) but there are at least a dozen operators with nationwide networks. These companies attract substantial business from abroad. In 1987 Britain imported over 50,000 tonnes of hazardous waste, twice as much as in the previous year, and another 130,000 tonnes of less dangerous waste for burial in landfills. Although small in quantity relative to the annual quantity of hazardous waste arising in Britain (said to be of the order of 500 million tonnes) the handling of imported waste is lucrative. The business is constantly criticized, but the environmental lobbies have not as yet succeeded in getting such imports banned.

Britain has no coherent waste policy, and the respective responsibilities of the local authorities and central government are not clearly defined. Some attempts are being made to popularize bottle banks and ensure the recovery of waste paper, but the public fails to display enthusiasm for such schemes. It is said that Britain produces some 20 million tonnes of domestic waste per year, rich in plastics, paper, glass and textiles. Hardly any of this is recycled, and very little is incinerated – which would at least provide energy. Another 400 million tonnes per year is said to come from industry, agriculture and mining, so there is scope enough.[9] The number of bottle banks has grown from a mere 77 in 1977 to more than 2,500 by 1985, and the scheme is said to be profitable, but the recycling of glass in Britain still amounts to only 2.2 kg per capita per year: in Switzerland it is 10 times this – 21.7 kg. These figures demonstrate the tremendous potential that exists in recycling. Material recycled without problem elsewhere, such as drinks cans, paper and board, lubricating oil and plastic, is hardly recycled at all in Britain.

However, the situation is improving and instances of effective waste recycling are increasing in Britain. Since 1979 East Sussex County Council has experimented with solid fuel made from rubbish in a pilot plant at Eastbourne. Some 20,000 tonnes of rubbish per year is shredded, extruded and dried to produce 5,000 tonnes of pellets, which are used to heat council buildings and are also sold to private companies for use by market gardeners. It is now planned to extend the scheme by building further plants of a similar size for the Isle of Wight and Hastings. This is encouraging, but there is certainly still a long way to go.

Ethylene plants are in use worldwide, and recycling has been used to great advantage, not only in recovering usable compounds, but in reducing production costs as well.[10] Off-specification products produced during start-up, shutdown and during plant operation upsets, which used to be flared, can be recycled through the process train, as illustrated in Figure 13.1. Savings of as much as US$250,000 can be achieved per start-up. This, and numerous other aspects of waste treatment and pollution control in the petrochemical industry, are admirably dealt with in a recent book by M. B. Borup and J. Middlebrooks.[11] In the petrochemical industry, as in any other industry, one must start with a complete survey of the potential waste, not only during production but also during start-up and shutdown. The information to be collected should include the characterization and volume of each potential waste, with rates of flow and their physical, chemical and biological characteristics.

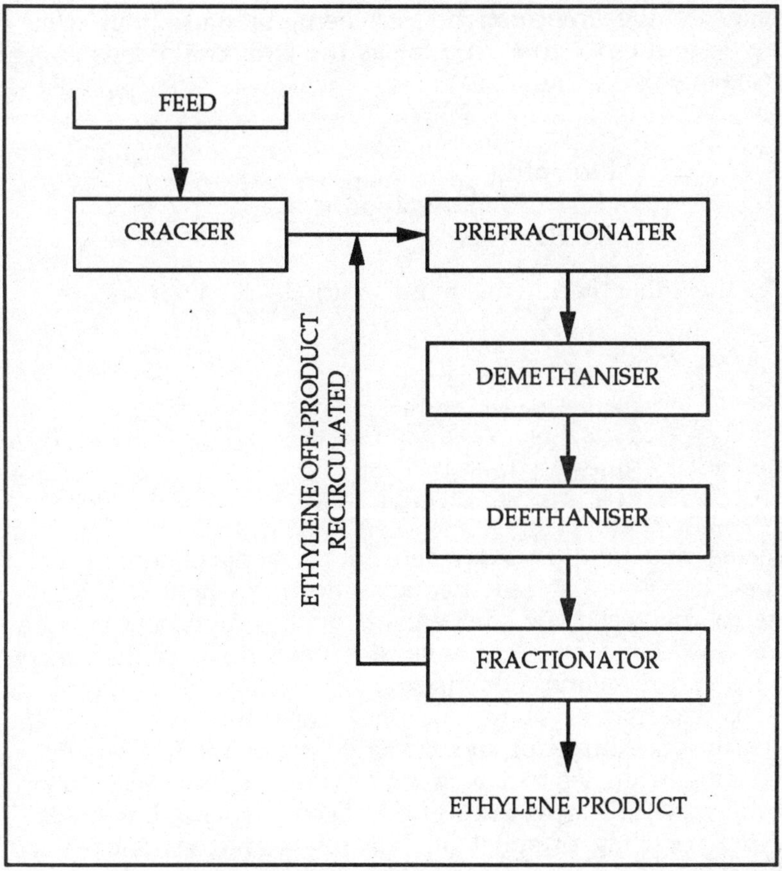

Figure 13.1 Recycling off-specification products. A simplified block flowsheet showing the way in which waste products should be recycled during start-up, shutdown and plant upsets.

Recycling spent oil

Spent lubricating oil is a messy pollutant, yet it can be converted into valuable products. Many car owners, especially in the developed countries, change the lubricating oil in their vehicles themselves, and they do this at least once a year. It is estimated that a mere 10 per cent of this spent oil is recycled, the bulk of the rest being dispersed in the environment, either by burning or by finding its way into surface water and landfill. Yet here is a valuable resource which can be recycled economically without damage to the environment.[12] Where does used oil come from and where does it go to? So far as the United States is concerned, the major sources are:

	%
Automotive	57
Industry and Aviation	36
Miscellaneous	7

On the other hand, the main methods of disposal are:

	%
Poured into ground	40
Placed with rubbish	20
Recycled	14
Burnt at home	4
Miscellaneous (unknown)	22

It seems that here, as elsewhere in the waste management field, Europe has given a lead. Perhaps encouragement has been given there to the recycling of spent oil by the high cost of imported crude. Fiscal mechanisms have also been devised that encourage widespread recycling. For instance, a tax on lubricating oil is used to subsidize its recycling, and this has resulted in the collection of nearly two-thirds of the available spent oil for re-refining. In fact, this scheme has had such remarkable success in West Germany that the tax on lubricating oil has been dispensed with. It seems that the recycling of spent oil can now stand on its own without subsidy.

Similar results have yet to be achieved in the United States and elsewhere, apparently due to lack of economic viability: it is more attractive to burn waste oil than to re-refine it. A plant to re-refine waste oil costs about four times more than a plant of the same capacity to process waste oil for use as fuel, but it does not bring in four times the profit. Figure 13.2 presents a block flowsheet of the refining process for used lubricating oil. A three-stage distillation process is required. First, water is removed: this requires little treatment before disposal. Next, the light hydrocarbons are removed: these are usually burned as fuel in the reprocessing plant. Finally, the oil requires a hydrogen catalytic treatment or clay filtering. The hydrogen process is more expensive, but it has the advantage of producing very little hazardous waste. With clay filtering, however, an oily clay eventually has to be disposed of as waste.

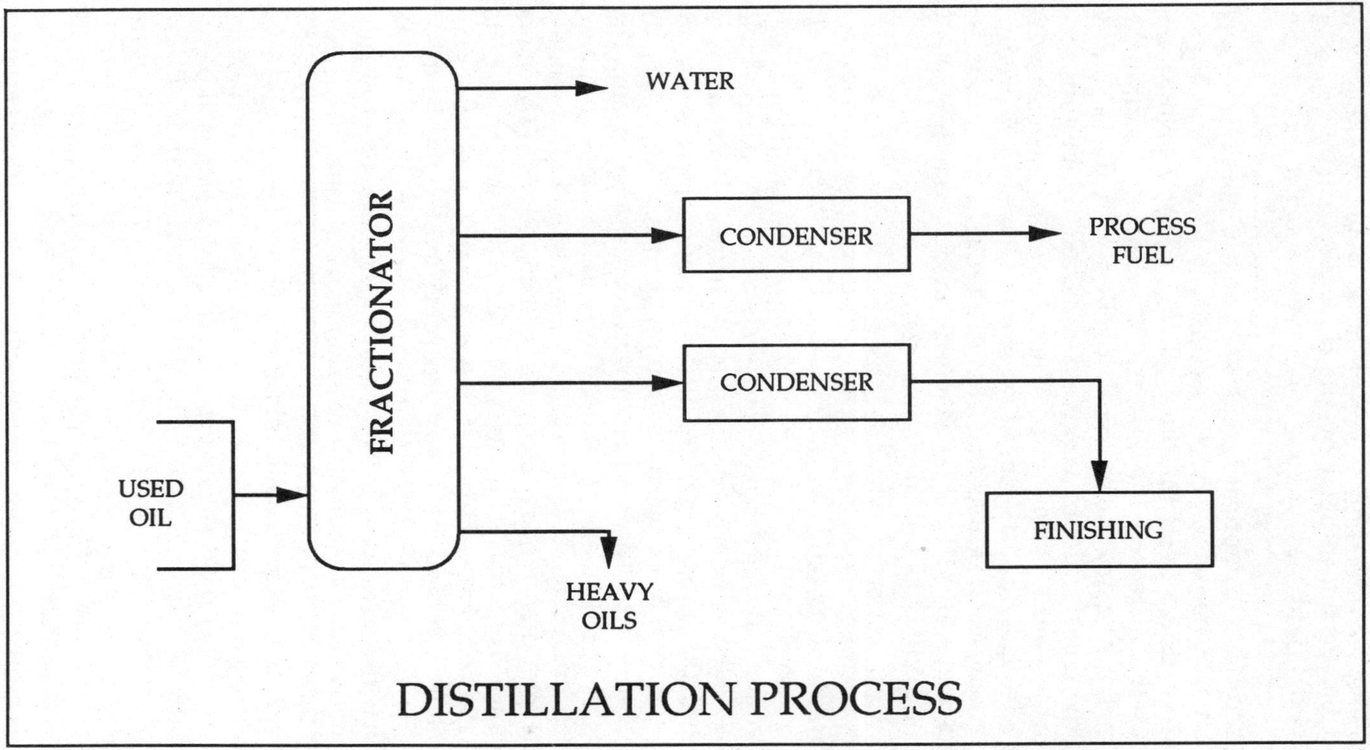

Figure 13.2 **Recycling spent oil. This block diagram illustrates the three-stage distillation process commonly used for the re-refining of lubricating oil.**

Nonetheless, the recycling of spent oil was once prevalent in the United States. Encouraged by the potential scarcity of oil, as in the Second World War, the industry grew and prospered until the 1960s, when a change in technology slowed the industry down severely. Additives to lubricating oils for cars, originally minimal, grew to some 25 per cent. Their purpose was to improve engine performance and to extend the life of the oil before a change was needed, but the increased life resulted in more sludge and dirt being present in the spent oil which, together with the additives, made its refining far more complex – indeed, a sophisticated chemical process was then required. Simple refining systems were no longer adequate, and the specifications for the refined product became ever stricter. Products that contained even a small amount of recycled material had to be labelled 'made from previously used oil', and this was a further disincentive.

The problem of plastic waste

The most commonly used plastic materials are not biodegradable and this has been a cause for concern. With the advent of what we might call the 'plastic age' and an exponential increase in the use of a wide variety of plastic materials, their safe disposal has presented a growing and, it seems, insurmountable problem. The incineration of plastic waste produces toxic fumes, and dumping them on a landfill results in vast accumulations of non-biodegradable waste. Attempts to encourage users to segregate plastic waste have met with limited success and, in any event, the problem of its ultimate disposal remains. However, it seems that, thanks to human ingenuity, some solutions by way of recycling may be at hand.

One of the biggest obstacles to the recycling of plastic waste has been that the various types of plastic waste cannot be mixed with one another: the chemical reactions that ensue cause too many processing difficulties. However, techniques have now been developed whereby incompatible plastics can be blended and still turned into useful products. A Belgian company, Advanced Recycling Technology, has developed an extruder which will blend a wide range of plastic materials such as polyvinyl chloride (PVC), polyethylene terphthalate (PET) and polystyrene (PS). The mix can then be formed into stable moulded products.[13] Over a dozen such plants based on this extruder technology are currently in use across Europe and in the United States, and the process is outlined in Figure 13.3. The extruder melts the mixture of plastics together, and this can then be moulded as appropriate. The product has wood-like properties; it can be nailed, screwed into, cut and planed with standard woodworking equipment. It is indeed better than wood in many applications, since it is water-resistant, rot- and bacteria-proof, and resistant to salt water and chemicals. Indeed, the very properties which make it so difficult to dispose of are of great value in re-use. The material will not splinter or split, withstands freezing and thawing and can absorb shocks. One use has been in the manufacture of fencing posts, and it seems that posts placed in the ground five years ago have remained upright and rot-free, requiring no maintenance.

The waste plastic cannot be accepted quite at random. It seems that the feed should contain some 60 per cent polyolefins, but whilst the rest can well be other plastics, non-plastic waste, such as paper and glass, as commonly found in municipal waste, can be accommodated to some degree. No outside heat source is used, since this can degrade the material's physical properties. The waste mixture softens under pressure as it is forced through the extruder by a high shear mechanical screw, whilst the short residence time in the extruder ensures that no volatiles are released to pollute the atmosphere. The polyolefin fraction softens and becomes the carrier of the other materials in the mix, which then act as fillers, lending rigidity to the end product. Since there is no chemical change during processing, the final specification of the finished product is determined by the nature and mix of the input waste. Since this cannot be controlled with precision, given its provenance, the nature of the product – in particular its quality and colour – is determined by the skill of the plant operator in achieving some degree of uniformity. The batches and the rate of extrusion have to be adjusted in a manner determined by the experience of those running the plant. But there is no doubt that this is a very acceptable way to recycle plastic waste, and it should eventually make a substantial contribution towards minimizing this particular problem. Vehicle manufacturers are also looking to recycle the plastic materials incorporated into their products. The plastic waste from this source (excluding rubber) is expected to double over the ten years from 1985 to 1995, but the metals now being replaced by plastic had one big advantage: there was complete recycling. There is no doubt that the chemical industry will have to develop a similar 'closed loop' system in relation to its plastic products. This aspect is seen as a major obstacle to the increasing use of plastic products by many vehicle manufacturers.

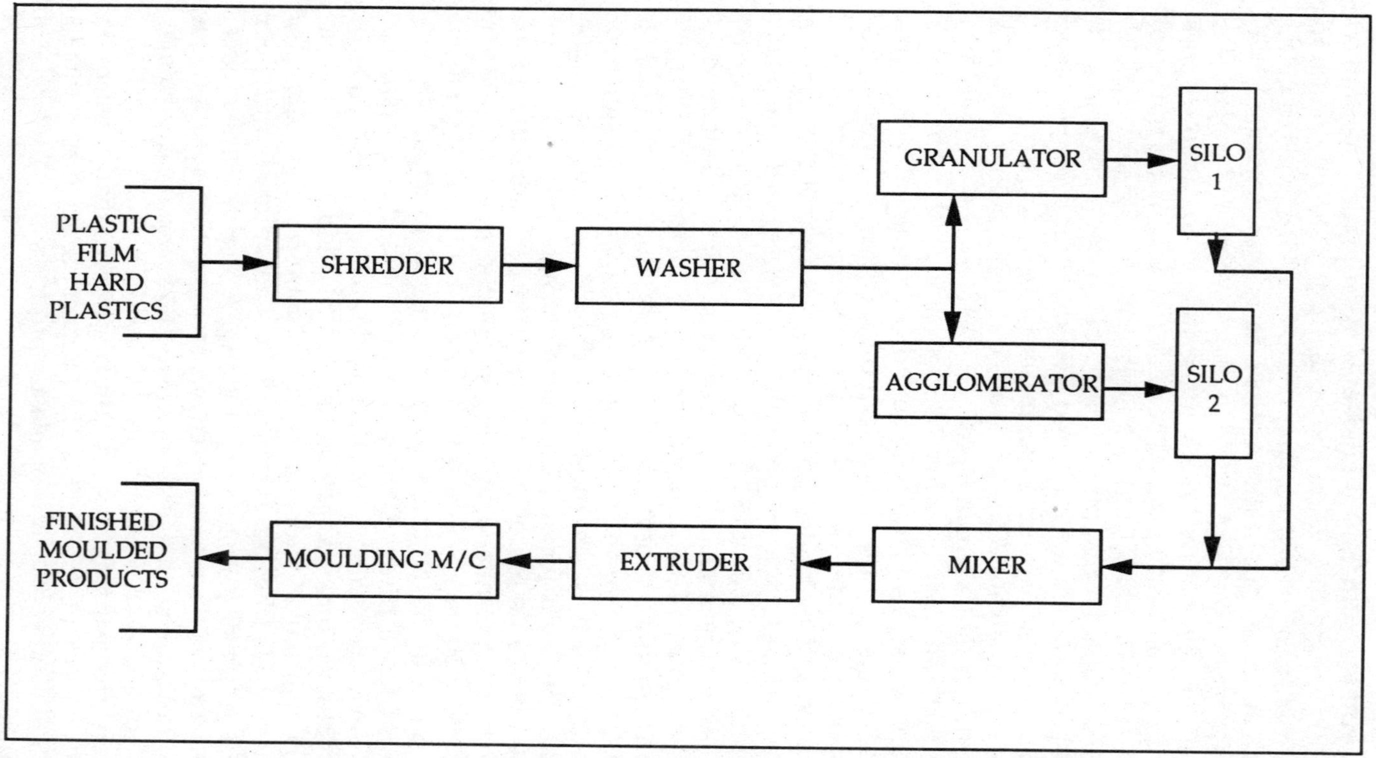

Figure 13.3 Processing waste plastic material. This diagram outlines the main steps in the recylcling of waste plastic material to produce finished moulded products.

Phenolic waste from lignite processing

As we have already made clear, the problems of waste disposal are not confined to developed countries. It is also a growing problem in developing countries. In this context, the experience of the Neyveli Lignite Corporation (NLC),, a public sector organization in India, is of interest. This corporation mines lignite, a fuel with characteristics generally similar to coal – indeed, it is often referred to as 'brown coal'. Their operations have produced a number of waste products which until now have had no use and have just been stacked away. But intensive research and development has resulted in the conversion of one such waste into a useful product. The lignite is carbonized to produce gas, which is then cleaned in a tar acid recovery plant. The effluent from this plant carries several complex polyvalent phenols, which is therefore called multivalent phenol (MVP). This effluent is a highly acidic, very complex material, without a definite composition. Even advanced analytical techniques have failed to identify the precise nature of this phenolic residue. In any event, its characteristics were always changing due, no doubt, to the high degree of variability inevitable in the original raw material, lignite, which is used as it is dug from the ground. It was this tremendous variability that hampered the finding of a suitable use for this phenolic waste material.[14] A number of possibilities were studied in depth, but without result. If it was fractionally distilled, some of the resulting chemicals would have no ready market. Used as a fuel, this tar-like material was found to be highly corrosive, spoiling the burners very quickly. In any event, such a solution was not really acceptable since, because of its phenolic content, the material was considered too valuable to burn. Biological degradation also had its problems: not only was the process expensive, but it posed a health hazard due to the residual phenols which would contaminate the ground water. So, facing up to their inability to dispose of this waste, which was being produced at a rate of about five tonnes per month, NCL stored the material away in drums, waiting for a suitable outlet. However, because of the corrosive nature of the material, the drums starting leaking, and a huge storage tank had to be built to contain the material.

Finally, development work by Indian Plastics and Chemicals (IPC) produced an acceptable solution, the process of which is outlined in Figure 13.4. After considerable experimentation, the company succeeded in developing a process which produced a wood adhesive, a wood preservative and an antioxidant for plastics and synthetic

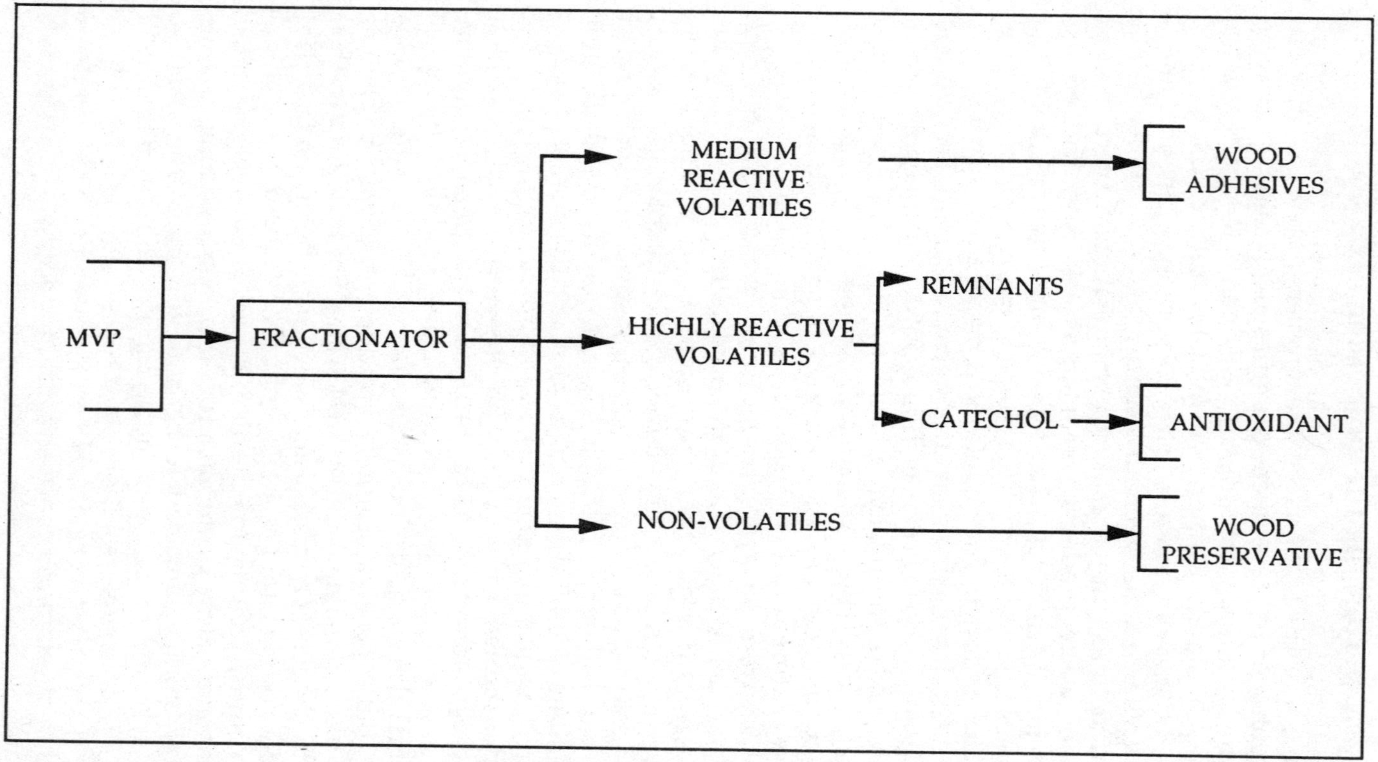

Figure 13.4 From phenolic effluent to useful products. This block diagram outlines the basic process steps whereby multivalent phenol is converted to produce wood adhesives, wood preservatives and antioxidant.

rubber, thus effectively disposing of the entire MVP waste. As Figure 13.4 illustrates, fractional distillation under carefully controlled operating conditions results in three major fractions, each of which was a commercial product having a ready outlet, and there were no residues. As a result, the entire accumulated stock of some 450 tonnes was purchased by the Indian Plywood Manufacturing Company Limited and is being successfully processed. That company also expect to lift at least part of the future production of this effluent. Adhesive resins suitable for different types of shuttering and boil-proof plywoods were manufactured, an end use that had the further beneficial result of not only high added value but also the conservation of valuable foreign exchange. Because the only phenol plant in India, run by Herdilla Chemicals, was apparently unable to meet the entire requirements of the entire plywood industry, IPM had a market advantage, being able, in times of shortage of adhesives when other manufacturers had to cut their output, to maintain a high level of production to satisfy not only the domestic demand but also cater for the export market. Thus the recycling of this phenolic waste had a very beneficial economic effect, as well as preventing further damage to the environment.

The examples we have given of recycling in practice, together with the many more that are cited in the literature, demonstrate very convincingly that recycling is a promising alternative for the safe disposal of waste, whilst at the same time providing useful products. Thus, not only are basic raw materials being conserved but the environment is also being protected. The steady exhaustion of existing resources and the escalating costs of energy have acted as incentives for the development of recycling technologies, but much more remains to be done. There is no doubt that future economic growth in many countries will draw heavily on recycling, and those countries who can make this transition quickly will not only grow faster but have a cleaner environment.

German inventiveness

The Veba Kraftwerke Ruhr AG (VKR) is a company operating in the Ruhr coal belt of West Germany, consuming over 7 million tonnes of coal per year in their power stations which have a generating capacity of 5,500 MWe. It is now compulsory in West Germany for flue gas desulphurization devices to be fitted to coal-fired boilers but VKR, having done this, has not only lowered the emissions of sulphur dioxide well below the prescribed limits, but has also lowered the suspended particulate matter and the emission of nitrous oxides. As a result, despite a tenfold increase in their electricity generation, the total emissions have remained practically constant. The company also recovers the flyash produced in the power plants, which is sold to the cement industry, with less than one-tenth going to landfill. The desulphurization plant yields gypsum which is sold for the manufacture of plaster board, such as is used for room partitioning. It is said that 30 per cent of the cost of a 700 MWe power plant now goes on environmental control. This represents a capital expenditure of some DM 350 million, with a recurring annual expenditure of DM 80 million.[15]

References

1 'Waste recycling: a two-way street', *Business India*, 26 January–8 February 1987, p. 11.
2 Barter Publishing Staff (ed.), *Recycling Commodity Exchange Encyclopedia*, Barter Publishing, USA, 1988.
3 Goldoftas, B., 'Recycling: coming of age', *Technology Review*, **290**, November–December 1987, pp. 28+ (9 pages).
4 Ibid.
5 Hershkowitz, A., 'Burning trash – how it could work', *Technology Review*, **90**, July 1987, pp. 26–34.
6 Rich, L.A., 'Solid waste – an overwelming problem offers business opportunities', *Chemical Week*, 27 July 1988, pp. 22–6.
7 Schwartz, J., 'Turning trash into hard cash', *Newsweek*, 14 March 1988, pp. 38–9.
8 'Britain's waste business – one man's poison', *Economist*, **308**, 3 September 1988, pp. 64–5.
9 Atkins, R., '2002 an ecological odyssey – laying waste to England's green and pleasant land', *Financial Times*, 14 January 1988.
10 Max, D.A. and Jones, S.T., 'Flareless ethylene plant', *Hydrocarbon Processing*, **62**, December 1983, pp. 89–90.
11 Borup, M.B. and Middlebrooks, J., *Pollution Control for the Petrochemicals Industry*, Lewis Publishers, USA, 1987.
12 Brinkman, D.W., 'Used oil: resource or pollutant?', *Technology Review*, **88**, July 1985, pp. 46+ (7 pages).
13 'Recycling mixed plastic wastes', *Industrial World*, June 1988, pp. C2+ (3 pp.)
14 Mruthyunjaya, H.C., 'Industrial waste utilisation', *Chemical Business* (India), 5–19 August 1988, pp. 52–4.
15 D'Monte, D., 'Environmental technology', *Business World*, 8–21 June 1988, pp. 72–3.

Chapter 14

RESOURCE RECOVERY: THE ULTIMATE GOAL

The previous three chapters have dealt with various aspects of waste minimization: first the need for the development of a non-waste technology; second, the principles and practice of recycling. Recycling may be an effective way of utilizing what would otherwise be wasted, but it is obviously even better if processes are so designed that there is no waste and hence no need for recycling. Non-waste technology, when applied, will ensure that waste is minimized if it cannot be eliminated, but existing resources may still be depleted. The absence of waste minimizes the use that is made of existing resources, but they are still being depleted. How can we recover the resources we use? This surely must be our ultimate goal. Is it possible?

An overview

Down the ages, Man has constantly sought to improve the quality of his life, whether at work, at home, or at play. Each generation learns from both the successes and failures of previous generations, and applies that learning. This is the heart of the message brought to us by that noted scientist, Bronowski, in his book *The Ascent of Man*. As he points out in his foreword:

> ...knowledge in general and science in particular does not consist of abstract but of man-made ideas, all the way from its beginnings to its modern and idiosyncratic models ... discoveries are made by men, not merely by minds ... this presents a philosophy rather than a history, and a philosophy of nature rather than of science ... .[1]

Through a continuous learning process, successive generations inherit the accumulated knowledge of previous generations. They stand, as it were, on the shoulders of their predecessors in their 'ascent', going ever higher, ever onwards and upwards. This is generally true in every field, and waste management is no exception to this. However, whilst increasing technology and innovation has obviously played a really vital role in the proper management of toxic and hazardous waste, it seems that many of the solutions so far have only produced yet another problem. This problem then demands further innovation and applied technology for its solution. Because of the way in which things have developed, most of the work in the field of waste management has been directed towards the handling, treatment and disposal of waste, not towards its minimization. This is unfortunate, since the minimization of waste should really be at the centre of all effort related to waste-handling. To check the creation of waste right at the start is obviously the most desirable solution, yet over the years it seems to have received little attention.

Non-waste technology follows this road, and should even question whether a particular product is needed at all. Systems analysis, or resource/energy analysis, is a powerful tool in the application of non-waste technology. It considers both the production process itself and the related environmental systems, developing a materials and energy balance, so that the net losses can be both evaluated and studied. All the various inputs and outputs relating to the manufacturing or fabricating process, including energy, equipment and investment have to be expressed in uniform monetary terms. This is difficult, but it is the only realistic and practical way to get to the root of the problem, and so learn what is actually happening. The strongest possible inducement to economy is to learn where 'the money is going' and how much is going down roads which lead to loss, and often to 'nowhere'.

Resource recovery economics

The most damaging situation in this respect relates to the production of toxic and hazardous waste, which not only involves direct loss but constitutes a risk which may eventually lead to further loss. It seems that resource recovery is the most cost-effective approach to toxic and hazardous waste management, and it is seen as sufficiently important to warrant a number of full-length books on various aspects of the subject. Typical of these is the work by Stuart Russell titled *Resource Recovery Economics*.[2] This work is based on the wide-ranging, practical experience of the author, who works as a consultant. A systems approach to feasibility studies is presented, illustrated by a hypothetical but nonetheless realistic case study.

The whole gamut of activities relating to the case study are covered, including analysis, resources, marketing, alternative systems, financial aspects including life cycle costs and then last, but not least, the implementation of the scheme. The approach adopted is a far cry from the usual wasteful approach to such problems. Although hailed as a 'cure-all' for waste disposal and energy conservation, resource recovery has yet to achieve widespread acceptance. Despite the thousands of studies that have been made worldwide, only a few have been translated into actual operating facilities. The lack of wide acceptance for resource recovery has been due to political, economic and even technical reasons, and perhaps even to the superficial nature of many of the studies. However, the pressure resulting from the two energy crises of 1973 and 1979, together with a growing concern about the steady exhaustion of precious raw materials – particularly fuels – and the mounting opposition by the general public to refuse dumps, landfill and incineration plants in their neighbourhood, has encouraged increased, and much more serious, attention to this subject. Resource recovery may not be the answer in every case, but it is essential that it be included as an additional tool when considering the options in relation to waste management. The 'systems approach' suggested by Russell, as illustrated by the case study, takes into account *all* the elements involved in waste management: transport, processing, combustion and final disposal. The approach, as demonstrated, also selects a series of 'system alternatives' and makes a comprehensive economic comparison before presenting the final decision. Some previously unpublished data from actual feasibility studies is also included to help in the making of realistic and comprehensive economic comparisons.

Systems analysis and life cycle costs

Once all the necessary basic data has been collected, including market identification and the selection of the appropriate system alternatives, the costs can be calculated and an economic analysis developed for management decision-making. For each of the possible alternatives it is necessary to decide:

1 the most economic location;
2 the most economic facility capacities and service areas;
3 the proper number of facilities in multi-facility systems.

Not only the initial capital cost but also the operating costs over the entire life of the project must be taken into account in order to obtain an objective, realistic overall cost comparison. The real objective for such a comparison is the total life cost, especially when several alternatives are being reviewed. With the appropriate data, even a sensitivity analysis can be carried out in order to assess the effects of changes in certain of the parameters affecting cost, such as inflation, interest rates, cost of energy and the like. Readers are referred to the work by Russell for the system and mathematical modelling, but the summary of the results of a cost analysis for a case study concerning the disposal of city refuse is given in Table 14.1. Whilst disposal by landfill appears at first

Table 14.1 Summary of a simple resource recovery case, relating to domestic refuse.

Alternative	Transfer	Processing	Revenue	Cost	Total Life Cycle Cost
1 Landfill (new)	5.7	4.7	–	10.4	163
2 Steam sold	3.9	16.6	8.8	11.7	86
3 Co-generation	3.9	20.2	9.3	14.8	107
4 Steam for district heating	3.8	17.6	5.4	16.0	170

Notes: 1 All costs are per ton, with the exception of the life cycle costs.
2 All figures are given in US dollars (1985).

Adapted from H. Alter, *Materials Recovery from Municipal Waste: unit operations in practice*, Marcel Dekker, USA, 1983

sight to be the cheapest alternative, this comparison ignores the future. Extending the analysis to assess the life cycle costs of the various alternatives reveals a very different picture. Based on certain assumptions in relation to rates of interest, alternatives 2 and 3 prove to be far the cheapest over the full life of the project. Alternative 2, since it is so favourable, needs to be pursued further: firm steam sales agreements would need to be negotiated.

Implementation planning

As we have just seen, Table 14.1 presents a summary of a simple resource recovery case, relating to domestic refuse, and indicates that a tentative decision should be made in favour of the scheme which offers steam for sale, perhaps to a local utility. Such projects need to be supported by firm and detailed plans if they are ever to be implemented via the bureaucratic systems that normally prevail in local government. The factors that provide justification should be highlighted. These are usually:

- economic factors, such as the low cost of disposal;
- offers least harm to the environment, in terms of pollution;
- uses a renewable energy source, and so conserves fuel, such as oil.

Unfavourable economics will, of course, mitigate against a resource recovery project, but there can also be legal barriers, or social or environmental issues that will delay implementation. These must be amended so that they not only encourage non-waste technology, resource recovery and recycling, but make them compulsory.

It seems that in the United States a number of resource recovery projects have failed either technically or economically. Technical failure has often resulted in financial failure. The reasons for failure have been many and various: there may have been a lack of proper planning; the wrong process may have been chosen; or there may have been unforeseen circumstances. The high failure rate seems to have instilled a 'fear of the unknown' amongst those contemplating such projects, with the result that there has been great hesitancy in implementing a project even when detailed cost analysis has shown that it is viable. Nevertheless, resource recovery is a preferred option and should be proceeded with whenever possible.

Some classic examples

By way of reassurance, perhaps we can cite a few classic examples of resource recovery. The iron and steel industry worldwide has been one of the most polluting of industries, giving rise to solid, liquid and gaseous wastes, all of which are obnoxious. However, new technology, such as direct reduction, continuous casting and scrap recycling, has made a substantial contribution to resource recovery, minimizing the impact of these plants on the environment. In the manufacture of metals in general, product-oriented closed circuit techniques need to be adopted, such as the Petersen process for alumina production, where iron and cement are the only byproducts, followed by smelting with total fluorine retention and subsequent regeneration of synthetic cryolite. The use of charcoal rather than coke in steelmaking, as is now taking place in parts of South East Asia and Japan, is another major improvement. Again, when copper ores are smelted with charcoal, steam and air, elemental sulphur is the byproduct, and this can be used for raising the fertility of sulphur-deficient soils.

In the case of non-metallic minerals, a major feature of their mining has been the total devastation of the countryside. This has occurred with the large-scale mining of chalk, limestone, gravel, sand and clay. But with imagination and commonsense, a drastic change has taken place. For instance, the Ready Mixed Concrete Company has converted the scarred Thames Valley, with its ugly waterlogged holes left by years of digging for the gravel used as concrete aggregate, into a superb recreational centre. Not only is this of great benefit to all those living in the area, but it is a profitable venture in its own right. The water sports complex that was developed became the site of the World Water Ski Championships in 1975. English Clays offer us another example in the same field. This company is a world-famous producer of high-grade china clay, and it is now utilizing the ever-mounting quantities of mine waste, which were threatening the surrounding countryside in Cornwall, for the construction of prefabricated housing – the Cornish Unit House. Even the spoil heaps that have grown up near deep coal mines have been transformed into recreational areas. We have the 'Wigan Alps' in the old coal-mining areas of industrial Lancashire in the UK, and the ski slopes on Mount Trasmore, in the United States.

Proper waste management

Resource recovery is an integral part of proper waste management. Municipal and industrial waste are a valuable resource and to just dump such material is sheer waste, apart from the pollution and the nuisance that can be created. Typical municipal solid waste contains not only domestic rubbish, but the waste produced by commercial activity and light industry, and scrap from construction

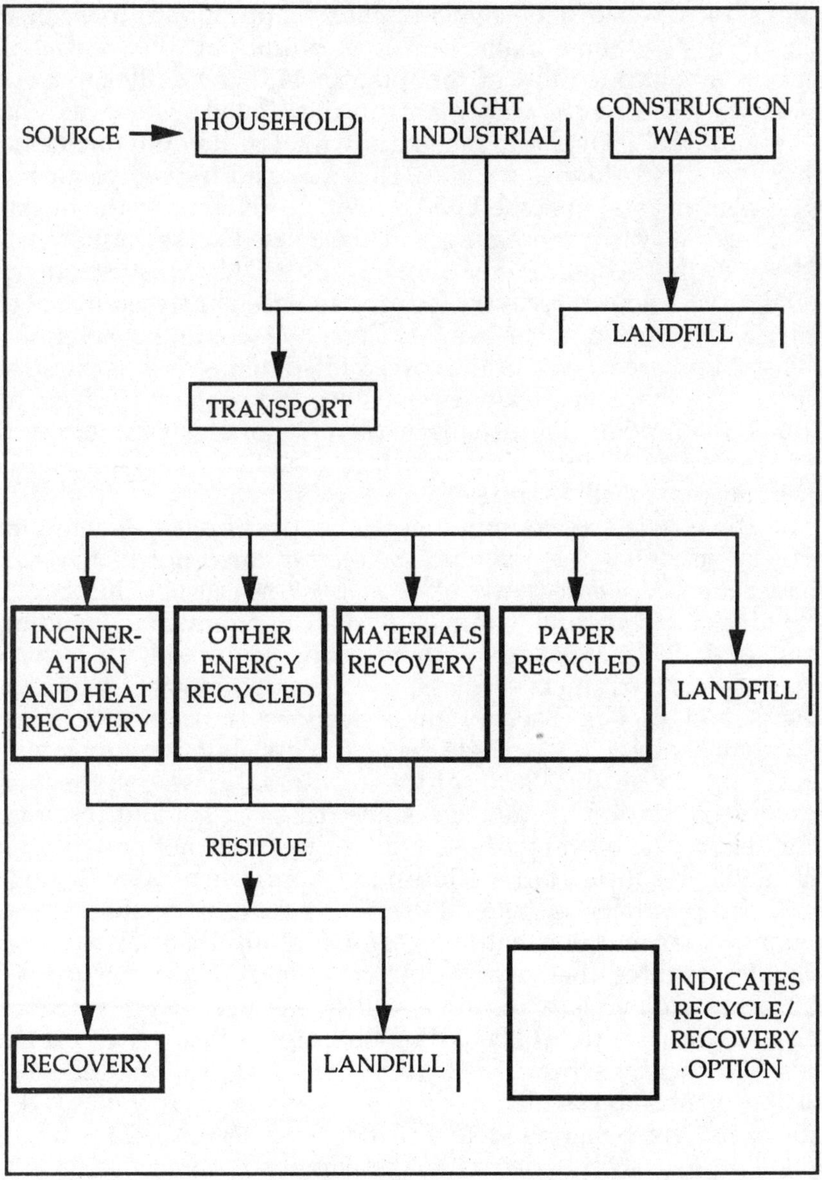

Figure 14.1 Managing municipal solid waste. Municipal solid waste has three major sources, as indicated, and it will be seen that, for the most part, such waste can be either recovered or recycled.

and demolition work. Figure 14.1 illustrates the possible recycling options in a waste management scheme appropriate to such circumstances. The three main sources of municipal solid waste each comprise about one-third of the total. In 1981 the daily capacity of municipal resource recovery plants in the United States was stated to be about 27,000 tonnes per day, with the amount incinerated being some 1,400 tonnes per day. This was said to involve a capital investment in total of some US$1.5 billion, exclusive of the facilities for industrial waste processing and the related scrap industry. The costs and the economic risks involved in the construction and operation of resource recovery plants have been analysed by several writers, the work of Alter being typical.[3] We would, however, warn against blind acceptance of the costs quoted in the various published sources. Comparison is dangerous, since the scope can differ, and various assumptions are usually made but not always stated in the text. One needs to do one's own estimate and cost analysis in order to arrive at a reliable figure.

For those wishing to investigate the possibilities for resource recovery in detail the reference work *par excellence* is the Kirk–Othmer *Concise Encyclopedia of Chemical Technology*.[4] This book is a condensation in one volume of their monumental 26-volume work, and deals at length with recycling and resource recovery. It covers ferrous metals, glass, non-ferrous metals, oil, paper, plastics and rubber. Each of these sections is dealt with by the respective experts in a single page or less, but we are given a concise picture of the 'state of the art'. In all these cases resource recovery is desirable, and in some cases it should be made compulsory. The burning of used oil, for instance, not only destroys a valuable resource, but pollutes the atmosphere. On the other hand, scrap rubber is better burnt as a fuel, since that is more economic than burning natural gas or fuel oil, themselves a more valuable resource than waste rubber. Indeed, when viewed as a fuel, scrap rubber is a resource which has been largely ignored. It has a heat value (14,000 BTU/lb) higher than that of most coals, which range from 8,000 to 12,000 BTU/lb. Shredded tyres can be burnt successfully in stoker-fired boilers. In Japan whole scrap tyres are being used to fire Portland cement kilns, but the potential in this area seems to be largely ignored. The United States alone generates some 250 million scrap tyres per year, but the bulk of this scrap rubber – some 75 per cent – is disposed of by landfill. A further 20 per cent is retreaded and a mere 5 per cent is reclaimed or used as fuel.

Fortunately, recycling and resource recovery is now attracting increasing attention from all those concerned with waste management, and the literature on the subject is growing fast. Conferences and seminars are being held at regular intervals, and the papers presented are published later and so made available to a wider audience.[5]

References

1 Bronowski, J., *The Ascent of Man*, British Broadcasting Corporation, London, 1973.

2 Russell, S., *Resource Recovery Economics – Methods for Feasibility Analysis*, Vol. 22 in series *Pollution Engineering and Technology*, Marcel Dekker, USA, 1982.

3 Alter, H., *Materials Recovery from Municipal Waste: unit operations in practice*, Marcel Dekker, USA, 1983.

4 Kirk, R.E. and Othmer, D.F., *Concise Encyclopedia of Chemical Technology*, John Wiley, New York, 1985.

5 Bridgwater, A.V. (ed.), 'Waste strategies for the future', *Resource Conservation*, **12** (3–4), October 1986, pp. 155–279; Bell, J.M. (ed.), *Proceedings of the 39th Industrial Waste Conference*, Butterworth, UK, 1985.

Part IV
THE ENVIRONMENTAL ISSUES

Chapter 15

POLLUTION ABATEMENT AND CONTROL

In dealing with waste disposal and treatment we have been looking at that which society has made that it does not want and therefore has to dispose of. But, of course, it would be far better if it were never made in the first place. We have seen that man-made waste pollutes the earth, the sea and the atmosphere so, to the extent that the production of this waste can be reduced, we shall have a cleaner world. In Chapter 2 we looked at one major pollutant, acid rain, examined it at length and noted the extensive damage that was caused. We also saw that the damage is international in character: it does not recognize national boundaries. There are a number of other results of pollution that are international in character – notably the depletion of the ozone layer and the 'greenhouse effect'. Pollution is very clearly global in nature, and now that the Soviet Union is emerging as a full participant in the study and resolution of the problems of pollution, worldwide action is beginning to develop in relation to its abatement and control. In our judgement, worldwide action is crucial to success.

It is one earth

The earth's vital functions have now deteriorated to such an extent that the 'patient' is clearly seen to be in declining health. What is more, the situation is going 'downhill' so fast that the policy-makers can ill afford to postpone the necessary investment now needed to implement an effective 'health-care' plan for the planet.[1] Those vital functions include:

- *Forest cover*: 31 million hectares damaged in the developed world, whilst the tropical forests are shrinking by 11 million hectares per year.

- *Lakes*: Thousands of lakes in the industrial northern hemisphere now biologically dead: thousands more are dying.
- *Topsoil on cropland*: 26 billion tonnes net loss annually.
- *Desert areas*: Six million hectares of new desert are being formed every year.
- *Fresh water*: The water table in parts of Africa, China, India and North America is falling as water withdrawal exceeds the recharge rates.
- *Plants and animals*: Thousands of species are becoming extinct every year.
- *Ground water*: Some 50 pesticides have been detected in ground water in 32 US states.
- *Climate*: The mean temperature of the earth may rise from 1.5 to 4.5°C over the next 50 years, with a consequent rise of some 1.4 to 2.2 metres in the sea level.
- *Ozone layer*: There is a growing 'hole', which suggests gradual global depletion. This could have disastrous results, as later discussed in Chapter 17.

A most significant report prepared by the World Commission on Environment, with the title *Our Common Future*, opens with the following words:

> The earth is one but the world is not. We all depend on one biosphere for sustaining our lives. Yet each community, each country, strives for survival and prosperity with little regard for its impact on others.[2]

Unfortunately this is very true, and we have already looked at a number of examples which illustrate this sad fact very clearly indeed. The fact that we live in one world and its implications must be urgently recognized by us all. We need to appreciate that we are interdependent. More than that: whilst we may think that by exporting our problem it is resolved, in the long run it will rebound upon us, as well as others. Everyone, rich or poor, has everything to lose if the peoples of this world continue along the road they are following at the moment, for it can only lead to destruction. This is true of many aspects of life today, but it is particularly true of pollution. A step in the right direction has been taken by what is known as the Montreal Accord of 1987, in relation to ozone protection, but that is a very small step along a very long road. It seems that very similar problems in relation to pollution are actually facing all the countries of the world: these problems are very evidently universal in character and they can only be resolved by joint action. A start has indeed been made with respect to ozone depletion, acid rain and the 'greenhouse effect'. It seems that the global climate is changing for the worse and that 'the earth's alarm bells are ringing'.[3]

Creating a sustainable future

Energy is essential to all industrial development, but it is the development of energy sources that has created one of the most intractable problems that confront the world today – a problem that is steadily growing more critical. The burning of fossil fuels creates carbon dioxide, and it is this ever-increasing volume of carbon dioxide in the atmosphere that is bringing the problems. It is indeed a two-edged sword. Not only are the most commonly used fuels of today damaging the lakes, the forests and the public health, but they are non-renewable. So what happens as they become exhausted? Nuclear energy is seen as a viable alternative, but is it an acceptable one? This is a matter for much public debate. What is certain is that the present energy system is polluting the atmosphere by as much as a tonne of carbon per person per year, and the continuing accumulation is believed to have already initiated a deterioration in both the climate and the general environment. The International Energy Agency in their Annual Report for 1987 has, for the first time, made some reference to the greenhouse effect, although they do not seem to see it as a matter for concern. We, however, see it as a matter of real concern – a subject that one dare not ignore.

The issue goes much deeper than the pollution of the atmosphere, since it involves a basic energy issue: the efficient, economic use of fuel. Fossil fuels are the main source of energy at the moment, and they also happen to be the main source of air pollution. Both the cities and the rural areas are suffering, and one result may well be the increasing incidence of respiratory and cardiac disease. The treatment of such disease is costly, so on a purely economic basis action should be taken, even if one were to disregard the suffering that is caused – suffering which cannot, of course, be priced. Yet, despite clear evidence that air pollution is bringing much damage to health in its wake, the European Community has not been able to agree upon truly effective standards for emission reduction. Many other countries have not yet even made a start in this direction. The crux of the argument is energy efficiency, and an increase in efficiency not only has its own pay-off, bringing lower costs, but it also brings with it lower pollution levels. It seems that inefficient combustion is the major contributory factor, and that a mere 1.5 per cent improvement in combustion efficiency could transform the situation, keeping the carbon dioxide concentration at a somewhat safer level.

Whilst the reduction of carbon emission through industrial activity must be pursued vigorously, an alternative, or rather parallel, solution is to plant trees on a massive scale. This would not only satisfy the constantly increasing demands for fuelwood, construction timber and paper, but would stabilize the soil and help restore the balance in the carbon cycle. It is this aspect that immediately concerns us. The trees (indeed, all vegetation) transfer carbon from the atmosphere to the terrestial system, at the same time releasing oxygen into the atmosphere. But to be meaningful, such action has to be taken worldwide, and the present wholesale destruction of the tropical rain forests would have to stop. It is said that the wholesale planting of trees could reduce the carbon dioxide content in the atmosphere very substantially, perhaps by as much as 30–40 per cent. The replacement of fossil fuel-based power by nuclear power would also reduce carbon emission, but this course is meeting with much opposition from the environmentalists who fear that one type of pollution is only being replaced by another, perhaps an even more dangerous, one. Unfortunately, the electrical power utilities are largely a monopoly industry and are not therefore under any pressure to improve the efficiency of their operations. If it were possible to throw open power generation to open competitive bidding, creating a free market, the situation could change radically. The industry would be compelled to increase their efficiency of operation for mere survival amidst keen competition. However, this does not seem to be a practical possibility. Fiscal incentives are another possibility, but no one country is likely to act on its own. Concerted global action is the real answer, but will it ever come?

Averting catastrophe

Can catastrophe be averted? Some think that it can, and J. G. Morone and E. J. Woodhouse have recently published a book entitled *Averting Catastrophe*.[4] Significantly, the subtitle is 'Strategies for regulating risky technologies'. The authors, one from industry

(General Electric) and the other from academia (Renselaer Polytechnic) discuss five separate technologies, which are dealt with in the context of the situation in the United States, although they seem to be of universal application. Each of these five technologies is seen to pose a potentially severe threat to the community. At least two of these, namely ozone depletion and the greenhouse climatic effect, most certainly cross national boundaries. We believe that the other three technologies discussed by the authors also have global consequences, so we propose to look at them briefly.

Nuclear technology is the third to come under review. We have already referred to nuclear power as a possible alternative source of energy, but the Chernobyl accident in 1986, despite being very patently a 'one-off', has vastly increased public opposition to this particular technology. Judging by the radiation levels later measured not only in neighbouring countries, but those far away, such as the United States, this particular accident most certainly had global consequences, the full effects of which may not be known for several decades. But is this sufficient to justify the abandonment of nuclear power as a source of energy? Is that the way to avert catastrophe? Whilst nuclear power has lost considerable credibility as a result of Chernobyl, many nations, including the Soviet Union, France and Japan, are still vigorously pursuing its development, and it may yet prove a more acceptable solution to the pollution problem than the use of fossil fuels.

The fourth risky technology concerns recombinant DNA. This is very different to the previous technologies, but it is still global in its impact. The technologies that we have just been looking at are subject to much debate in the scientific community, but they seem pretty well united on the risks that are presented by DNA research, and further that the benefits in the agricultural and medical fields far outweigh the risks involved. Once again, pollution is the real risk. Some of the risks have been identified and analysed: these include the survival of organisms outside the laboratory or factory; the creation and release of dangerous organisms; and transmission to other organisms. The risks are there, but it seems to be agreed that the risk can be contained. This belief prevails despite the fact that the risk is invisible. Yet with nuclear power, which presents another equally invisible risk, fear is prevalent. Part of the problem seems to be the reaction of the lay person to the possibilities. Nuclear power, it seems, implies awesome disaster to the lay person, a prospect inspired by the consequences of the atomic bombs dropped at Hiroshima and Nagasaki. These fears have been reinforced by the accidents at

Three Mile Island and Chernobyl. But, as yet, there has been no mass media scare generated by biotechnology and bioengineering. The dire warnings of some fiction writers have failed to arouse the public imagination and create an atmosphere of fear. Further, the building of the necessary laboratories and factories is on a relatively small scale, and has attracted far less public attention than the construction of nuclear power stations. But the pollution risk could be serious: there is a need for rigorous supervision and very comprehensive containment techniques.

The fifth technology relates to our present subject, namely the containment of the adverse effects of toxic chemicals, such as pesticides. Prohibition is not seen to be a practical solution, so the possibilities of limitation and elimination are reviewed. This is one field where the lessons seem to have been learnt largely by the costly and time-consuming process of trial and error. This means that the processes of legislation and regulation have been slow and often inadequate. Of the many thousands of dangerous chemicals, it is said that a mere 0.4 per cent have been banned because they present an unreasonable risk. This low rate is ascribed to:

1 voluntary restraint by producers, doubting whether approval would be forthcoming;
2 the original estimate of the number of dangerous chemicals which was far too high;
3 the procedures currently adopted not being capable of screening out some of the dangerous chemicals.

In addition, one must not forget the political pressures created by chemical producers and their associations, seeking to preserve their commercial viability.

Absolute liability

The Bhopal disaster in December 1984 alerted the public worldwide to the risks they could run from hazardous chemicals released into their environment. As a result both of public pressure and the real concern of manufacturers and legislators, the toxic waste laws both in the United States and elsewhere have been strengthened to require chemical producers to report on any of the 300 or so hazardous substances listed, if they are used, stored or released into the atmosphere.[5] The next step, of course, is to regulate and reduce the risk resulting from these chemicals. Here the onus has been put on the producers themselves – rightly so. If a product poses a risk to the health of consumers it must carry a proper warning, failing which the producer can be sued for damages. This self-enforcement scheme was the brainchild of specialists from powerful environmental groups such as the Sierra Club and the Environmental Defense Fund. It is to be hoped that the same principle is eventually applied worldwide.

Of the listed hazardous chemicals, attention was first focused on the more dangerous, said to number some 30. These, when sold to the public, have to carry a warning as to the risks they present, such as cancer or birth defects. The ultimate objective is, of course, to reduce or eliminate the use of hazardous chemicals voluntarily. This may well result in companies being held liable for compensation, once the implications of placing all responsibility for the consequences squarely on the producer have been recognized in line with a notable decision by a Supreme Court judgment in India, which held unanimously (5–0) that:

> ...enterprises engaged in a hazardous and inherently dangerous industrial activity are *strictly and absolutely liable* to compensate the victims of any accident caused by the activity.[6]

The emphasis is of course ours. The Court also commented on the degree of compensation that should be paid, saying that it must have a deterrent effect and should therefore be related to the magnitude and the capacity of the enterprise. Thus:

> The larger and more prosperous the enterprise, the greater must be the amount of compensation payable to the victim.

We shall take up this aspect of pollution prevention again later, when we come to consider the role of management in this matter.

The list of toxic substances includes two contentious items: tobacco and alcohol. For several years now, tobacco sold in the United States, the United Kingdom and a few other countries has been carrying the warning that it is 'dangerous to health'. In the case of alcohol, from the beginning of October 1988 in the United States all shops selling beer, wine or spirits are required to warn buyers that alcohol can cause birth defects. It seems that the protection of the unborn is a strong motivation in relation to legislation against the damage that can be caused by toxic chemicals. Once again in the United States, the supplier is required to state that no 'observable effect' follows exposure to a toxic chemical in the product being sold, even if it were to be a thousand times the level actually to be found in the product. A lead in this respect was actually given by the State of California, where a certain Proposition 65 was approved through the public voting system in 1986. The objective of this particular proposition was to eliminate, or at least reduce, the risks to health presented by toxic substances. Thanks to a powerful campaign in the local press it was voted in by a large majority, despite the opposition of industry and certain powerful business interests.

So much for the problem of poisoning ourselves, but what about poisoning others? We do not propose to discuss here what is called 'passive smoking', the inhalation of tobacco fumes from others. Let us look rather at a classic case of this sort – the poisoning of the Rhine in 1986,[7] which had serious and widespread international repercussions. Described at the time as Europe's worst toxic spill for a decade, an 'ecological disaster' and even 'a Bhopal on the Rhine', its effect may well last for years. It all began with a fire at a Sandoz warehouse in Schweizerhalle, just outside Basel, Switzerland. The fire started in the early morning of Saturday, 1 November 1986 and was attended to by the regular staff, since the specially trained chemical firefighters were off-duty for the night, although we do not think that this would have made any material difference to the end result, since no one really knew what health hazards were being posed by the fire. Later it was found that the main component in the smoke was phosphor-ester, a toxin. The water used for dousing the fire contained phosphorus-based chemicals and these, together with the pesticides and herbicides stored in the warehouse, were carried straight into the Rhine. This lethal brew also included the deadly ethyl-parathion, known to kill half those who absorb it even in small quantities. Perhaps an even more serious threat was posed by some 12 tonnes of a fungicide, ethoxyethyl, containing about two tonnes of pure mercury, stored in the warehouse at the time of the fire. At least some of this certainly found its way into the Rhine.

In all, about 30 tonnes of toxic chemicals were washed into the river and went on their way downstream. The Swiss officials were caught unawares, and reliable information was hard to come by. At the weekend, of course, all the offices were shut. In the West German town of Lorrach the police saw the flames shooting into the sky and radioed for information. The mayor of the nearby French town of Saint Louis, Theo Bachman, was not aware of the accident for nearly 12 hours. Swiss officials later admitted delaying an international alert for nearly 24 hours, because of some misunderstanding. Or was it perhaps just a 'big mistake', with those close to the incident not realising just how severe it actually was.

The result was tragic. Vast stretches of the Rhine were transformed into a river of death. There was total devastation along the first 300km of the river, from Basel to Mainz. Virtually all living organisms were destroyed, including perhaps half a million fish and 150,000 eels, whose bodies were thrown up as flotsam along the banks. As the full dimensions of the disaster came to be appreciated, there was much noisy protest, both against the presumed negligence of Sandoz (whose warehouse it was) and the slow response of the Swiss government. Later, Sandoz admitted it was at fault and formally pledged compensation for 'proven cases' of damage. The disaster was seen as comparable in magnitude to the nuclear disaster at Chernobyl and nicknamed 'Chernobale', Bâle being the French name for the town of Basel. Not only the public but the governments of France, West Germany, Luxembourg, Belgium and the Netherlands, all seriously affected by the disaster, blamed the Swiss authorities for allowing it to happen. At public meetings dead eels were thrown at Sandoz officials. There was a hidden irony, in that the Rhine had only recently become a symbol of ecological renaissance. It had been subjected to a massive clean-up operation and, after many years, the residents at Basel had been able once again to enjoy swimming in the Rhine. Some 10 varieties of fish were thriving, including the delicate trout, and the first of 30,000 salmon fry released in 1981 were about to reappear as disaster struck. Now the process must begin all over again.

This is all the more regrettable since Europe is not short of international conventions designed to be a guard against pollution. There are two covering protection of the Rhine, one for the Danube, others for the Baltic, the North Sea, the Mediterranean, cross-frontier air pollution, the protection of Europe's seas from pollution streaming off the land, and so on. These conventions embody various recommendations, but it seems there is no system of enforcement.[8]

Just as Bhopal has proved in some ways to be a 'blessing in disguise', so this incident on the banks of the Rhine may well have hidden benefits. It seems that Europe is now likely to be united as never before in its determination to tackle their common problem: pollution by toxic waste. The emergence and growth of strong, vigorous political groups, active in the cause of preservation of the environment, such as the 'Greens' in West Germany, seems to be having a beneficial effect.

The deadly PCBs

The technical abbreviation 'PCB' for 'polychlorinated biphenyls', which appears time and again in the text, must be familiar to many these days. PCB is a highly toxic poison that has been released into the environment over the years without control, because the dangers were not initially appreciated. It seems that this particular chemical has a long-term impact. Once it has found its way into rivers, lake and the ocean, it does not degrade but accumulates in fish and so enters our food chain. Indeed, it has been very properly described as the 'poison that won't go away'.[9] PCBs have been used in large quantities now for nearly 60 years. Initially they were used to replace mineral oil in capacitors and electrical transformers, as used in large numbers by the electrical utilities and railways. Highly stable and fire-resistant, they were seen as a boon to mankind. Now they are infamous. The first sign of trouble appeared in Japan in 1968, when over 1,000 Japanese were afflicted with 'oil disease' after consuming rice oil heavily contaminated with PCBs. The symptoms of the disease include acne, swollen limbs and liver disorders. Nevertheless, it seems that for over 30 years (1946–76) General Electric had been discharging PCBs into the Hudson River with water effluent from its capacitor plants at Hudson Falls and Fort Edward, New York State. In 1975 rock bass fish caught in the Hudson were found to contain 350 ppm of PCB, nearly 70 times the safe level of this contaminant as determined by the Food and Drug Administration. The Hudson was immediately closed to all fishing and General Electric was later found guilty and fined US$4 million. The New York Department of Environmental Conservation (DEC) voluntarily paid US$3 million for its negligence as a public 'watchdog'. These payments were used for PCB research and for a clean-up of the river. However, further research showed that what had been uncovered was but the 'tip of the iceberg'. It was later found that virtually every river in the USA had been contaminated with PCBs.

Despite this very serious situation, first disclosed over a decade ago, the EPA has still to finalize its approach to the problem. In a ruling made in 1985, it ordered utilities and landlords to remove several thousand 480-volt transformers from commercial buildings by October 1990. Other transformers must be registered with the local fire department, and combustible materials should never be stored adjacent to such transformers. Where transformers containing PCBs posed a threat to food or animal feed they had to be taken out of service forthwith. It is interesting to note that suspicion was first raised in this context as long ago as December 1977, when PCB was discovered in chicken feed at Country Pride Foods, El Dorado, Arkansas. The contamination was traced to a fishmeal warehouse in Ponce, Puerto Rico. It seems that, during a fire there in April 1977, PCB had leaked out of damaged electrical transformers and contaminated the fishmeal. Clearly the use of PCB in transformers presents a continuing hazard. The removal of PCB from existing transformers, allowing their reclassification, has emerged as a new and profitable business, since it costs less than 50 per cent of the cost of the new non-PCB transformer that would otherwise be required. But it seems this is a slow business and the market is not infinite.[10]

Part of the problem has been the difficulty in finding a suitable, safe substitute for PCB. Many early attempts in this direction failed,[11] and it seems that the options in this direction are very limited. In the absence of a suitable fluid to replace PCB in existing transformers, the only solution is to replace the transformer completely. Apart from the cost, the owner is still left with a problem: the existing transformer, particularly its PCB content, has still to be disposed of safely. Dumping it is *not*, as we have seen already, a satisfactory solution. More recently however, several companies have come forward with an acceptable option. For instance, Unison of Dublin, Ohio, a joint venture of Union Carbide and the McGraw Edison company, offer their *Reclass 50* as a refill medium and this has now been used for several hundred such transformers in the United States. A proprietory dialectric fluid is used in order to leach out the PCB from the transformer's internal windings and then replace it with a new dialectric. But the scrubbing and leaching process can take many months, the time taken depending upon the transformer size, the PCB concentration and the operating temperature. Even though the manufacture of PCBs has now been banned, it is estimated that, in the United States alone, there are some 300 million pounds of PCBs held in some 150,000 old transformers. Such old transformers as remain in operation are required by law to be inspected frequently, and are generally governed by very strict controls to ensure that there are no spills or leakage of the transformer fluid.

Unfortunately, this much feared chemical continues to make news not only in the United States, but also across Europe. A recent scandal concerns waste oil and involves Belgium, The Netherlands and Great Britain.[12] It came to light following a probe in the West Midlands (UK) during an investigation of firms engaged in recovering and recycling waste oil. Random checks revealed that the chlorine concentrations in the oil were as high as 21,000 ppm. For such material to be burned safely the proper equipment is required, with the appropriate emission controls, under the supervision of the proper local pollution authority. Estimates indicated that some 200,000 tonnes of this oil were on offer as heating oil across Great Britain, thus presenting a substantial threat to public health. It seems that oil contaminated with PCBs could be traced back to hundreds of different traders in Belgium and the Netherlands, who were shipping it to Great Britain for sale.

Thus, even though the manufacture of PCB has now been phased out, the problem is still with us. The danger presented by toxic chemicals remains, despite the increasing care that is now being taken. PCBs are only one example: their complete and safe removal is very difficult, as we have seen. Constant vigilance is required, not only on the part of the regulatory authorities, but also on the part of the public. This is where the organizations concerned with our environment can play a crucial, most important role. Meanwhile, the law in Britain and elsewhere is likely to become as strict as it now is in the United States with respect to the burning of PCB-containing materials. It is likely that there will be requirement for the so-called 'four nines': a destruction rate of 99.9999 per cent.[13]

References

1 Brown, L.R., Flavin, C. and Wolf, E.C., 'Earths's vital signs', *The Futurist*, **22**, July–August 1988, pp. 13–20.

2 Report: *Our Common Future*, World Commission on Environment, Oxford University Press, Oxford, 1987.

3 Smart, T., 'The earth's alarm bells are ringing', *Business Week*, 11 July 1988, p. 25.

4 Morone, J.G., and Woodhouse, E.J., *Averting Catastrophe*, University of California Press, 1986.

5 'Pollution – business beware', *Economist*, **306**, 20 February 1988, pp. 35–6.

6 Kharbanda, O.P. and Stallworthy, E.A. *Safety in the Chemical Industry*, Heinemann, London, 1988.

7 Hewitt, W., 'The poisoning of the Rhine', *Newsweek*, 24 November 1986, pp. 20–6.
8 'Europeans do it to each other', *Economist*, **301**, 15 November 1986, pp. 43–4.
9 Culhane, J., 'PCB's: the poison that won't go away', *Reader's Digest*, December 1986, pp. 112–16.
10 'PCB's make a slow exit', *Chemical Business*, September 1985 p. 72.
11 Brooks, K.W., 'Fewer headaches for PCB transformer owners', *Chemical Week*, 14 May 1988, pp. 33–4.
12 Milne, R., 'Inspectors uncover scandal of deadly heating oil', *New Scientist*, 24 July 1986, p. 19.
13 Rendell, J., 'On or off-site disposal?', *Process Engineering*, March 1988, pp. 71+ (2 pp.).

Chapter 16

ENVIRONMENTAL BIOTECHNOLOGY

It seems that most forms of waste treatment tend largely to defer problems for posterity rather than solve them, with the immediate result that the sea, the land and the air are being steadily polluted. In the process of time, this could have disastrous, possibly catastrophic, consequences. There is, however, one notable exception to this: biological treatment. The biological treatment of industrial, domestic and other waste presents little, if any, risk to the environment. Although there are some 'ifs and buts' hedging that statement, there is no doubt that the biological treatment of waste is extremely effective and innocuous. We shall seek to present the views for and against biological treatment, leaving it to our readers to decide to what extent this particular solution to the problem of waste disposal should be pursued.

The invisible allies

Environmental engineering has emerged as a specific discipline. The terminology first came into use in the 1960s, although its roots go far back in history.[1] The four major disciplines concerned in this area are: civil engineering, public health, ecology, and ethics. Environmental engineering has a proud history and a bright future. It is not only satisfying work, but profitable, challenging and enjoyable. Environmental engineers try to be part of the solution to a specific problem, whilst recognizing that *all* people are part of the problem as well.

Another, and perhaps even faster-growing, field in this area is that of bioenvironmental engineering, which is attracting people from various disciplines, including civil and chemical engineering, general biology, zoology, microbiology and chemistry. Problems are solved using an interdisciplinary approach. Environmental pollution is so serious and urgent a problem that its solution requires those specifically trained in this field. What is really required is a new breed of multidisciplinary engineers rather than a team of narrowly trained experts.[2]

Biological processing is an aspect of waste treatment that has attained sufficient importance to have given rise to a number of full-length books dealing with various aspects of the subject. Typical of these is a work by Winkler, *Biological Treatment of Waste Water*[3] and a volume, *Environmental Biotechnology*, edited by Foster *et al.*[4] The latter book has 12 chapters, contributed largely by British authors, although there is a chapter each contributed by authors from Australia and Switzerland. Biotechnology is a complex and highly technical subject, but to bring you a 'flavour' of this fascinating world of 'bugs', listed below are the subjects dealt with by the 12 authors:

- Aerobic processes: design criteria, operational and performance aspects
- Anaerobic wastewater treatment processes: biochemistry, digestion processes, digesters and reactors
- Mineral leaching with bacteria: chemistry and exploitation
- Composting and straw decomposition: process parameters
- Solid waste: processing, sludge disposal and utilisation
- Agricultural alternatives: probiotics, Rhizobium bacteria
- Microbial control of environmental pollution: genetic techniques, catabolic pladmids, recombinant DNA techniques
- Continuous culture of bacteria with special reference to activated sludge waste water treatment processes: bacterial growth requirements, batch and continuous reactors
- Cell immobilization: techniques, reactor design, performance.
- Aeration of waste water: basic physical factors, mass transfer, interfacial surface area, oxygenation rates
- Process engineering principles: material and energy balances, fluid flow, heat transfer and mass transfer
- Biopossibilities: the next few years, liquid and solid wastes, conclusions.

Biotechnology has been defined by the European Federation of Biotechnology as the integrated use of disciplines such as biochemistry, chemical engineering and microbiology, in order to develop the industrial application of microbes and cultured tissue cells. Manifestly we have to leave the process developments, with the design of the appropriate plant, to the experts. The book referred to above, despite its scope as we have detailed, is limited to the specific application of biotechnology for the management of environmental problems. It demonstrates the way in which the processes discussed can be integrated with non-biological technologies, which still have a part to play. Most of the significant activities in these fields are covered, with the control of pollution receiving continual emphasis. That is indeed where biotechnology comes into its own.

There is no doubt that biotechnology can play a significant role in both protecting and rehabilitating the environment, possibly even undoing damage which has been done in the past. The biological treatment of waste water is already a well proven technology, and is in fairly wide use. However, the processes currently available can only fully degrade carbonaceous compounds with a relatively simple structure, and ammoniacal compounds. Inorganics, toxic chemicals and structurally complex compounds can be bound to the biomass or degraded to some degree, but the extent of their removal is much less controllable. Thus, for instance, no guarantee can be given in relation to the removal of heavy metal ions by the activated sludge process. However, phosphates and nitrates can now be removed by biological techniques, and intensive work is in progress in many parts of the world. It is certainly the 'in' subject for research and development, not only in industry but in the academic world as well.

When it comes to solid wastes, one must question the wisdom of spreading metal-polluted sludge on agricultural land, even though this is permitted. The question remains: what will happen after 20 or 30 years? It would be more acceptable to process the sludge: a pretreatment for detoxification and metal recovery is indicated even though this would be very costly. There are known processes, whereby the metal ions are removed from the sludge by an acid treatment and then re-absorbed on specific bipolymers. With a better understanding of the metal/polymer binding reactions, it is possible that the process could be further developed. The polymers could then be so formulated that the absorption phase produces a material analogous to a low-grade ore, which could then be processed to recover the metals. Then we really are back where we began. Alternatively, and perhaps preferably, a biotechnological process could be developed that would stabilize or detoxify the sludge, but this would require specifically adapted or engineered microbes. It is not our intention to enter in detail into the complexities of this subject, but we do wish to demonstrate the possibilities.

Fortunately, not all solid waste contains toxic materials. Domestic refuse, for instance, is relatively innocuous and is admirably suited to bacterial decomposition. However this demands that the appropriate waste disposal policies be developed, based on reclamation (by biological treatment) and recycling rather than the present destructive technologies. As we have seen, that approach leaves the problem with us: it does not eliminate it. The technology already exists that would enable the chemistry of landfill decomposition to be so adjusted as to maximize biogas production, but it seems that little or nothing is being done – perhaps due to a lack of motivation. There are at least four proven technologies for large-scale composting, using mechanical mixing and aeration.[5] Perhaps by far the most economical and innovative process, holding great promise for the future, is that of vermicomposting. This involves composting with the aid of earthworms. Its unique feature is that mechanical mixing and aeration of the soil is provided by the earthworms themselves, which serve as bioreactors supplying essential enzymes. This leads to a highly efficient biodegradation of the waste materials by bacteria as it passes through the earthworm's tubular bioreactor.[6] In addition to the production of vermicompost, which is much superior to ordinary compost, there is also the production of the earthworm-biomass, which can be an excellent source of protein for fish, poultry, cattle and even for humans! Among the other possibilities presently under investigation are poultry waste treatment plants, production of vermicastings from sugarcane residues, a vermifilter unit to treat sewage and a vermicomposting plant based on the biodegradation of municipal solid waste. The possibilities really seem endless and it all started from some basic research by Charles Darwin more than a century ago. His paper 'Darwin on earthworms – the formation of vegetable mould through the action of worms', first published in 1881, has recently been reprinted to celebrate his centenary year. Many of Darwin's observations and findings have since been corroborated by extensive research and earthworms are now seen as most valuable creatures, with a special role to play in the recycling of waste to produce usable materials.[7]

Are there species other than earthworms that could be encouraged to play such a controlled, scavenging, degrading or decomposing role? Biomass cultivation is another possibility, coupled to the removal of the plant nutrients such as nitrogen and phosphorous from water: these are detrimental to the environment when present to excess. The continuing objective is to return the waste to the natural cycle: only in that way will the natural balance be restored and maintained, and in the process we will have much to gain. We are indebted for much of the above information to Mr Uday Bhawalkar, gained when we visited his earthworm farm in June 1989.

Bugs versus hazardous waste

When we look at the disposal of hazardous waste, it seems that the damage has already been done. A complete clean-up of the thousands of toxic waste sites scattered across the world could take 100 years and would cost a phenomenal amount. In the United States alone, according to an estimate prepared by the Office of Technology Assessment (OTA) such a clean-up could take 50 years and cost upwards of U$$100 billion. Who is prepared to pay? Unfortunately, such a high cost is inherent in the present methods of treatment, handling and disposal. What is more, the public is vehemently opposed to such work being carried out, at least ' not in my back yard' (NIMBY) and at their cost.

However, it is possible that mother earth can show us a better and cheaper way. On-site bioreclamation or bioremediation, as it is called, combining biological, chemical and physical processes has real promise. Not only would such an approach be much less costly, but it would also eliminate, or at least reduce, the pollution as well as the economic and legal liability that goes with the present conventional waste disposal projects.[8] But, despite its evident advantages, this technique takes much hard work and intensive study to implement, and it seems difficult to sell as a concept. The time taken to clean up the waste can be halved, but success demands a very careful balancing of the various environmental variables. Even though the technique is 'down to earth' and natural forces play a vital part, it is certainly not sufficient merely to throw the 'bugs' on the waste and then hope for the best.

It is said that biological treatment can cost less than half the cost of disposal by landfill, and only one-third of the cost of on-site incineration, yet it is little used. The technique has been used successfully for many years now, but it remains difficult to sell. John Biesz, vice-president of the Polybac Corporation, a subsidiary of the Cytox Corporation and specialists in producing adapted natural micro-organisms, suggests that the problem lies in the hidden nature of the processes involved:

> ...the biological cleanup has been a difficult concept to sell because few people charged with the cleaning up of hazardous wastes understand how seemingly invisible micro-organisms can live in an environment containing hazardous or toxic chemicals. They only understand something tangible, like bulldozers and incinerators.

He could have added 'big' to tangible: the 'bugs' are not only mysterious but very, very small. However, with more information and knowledge, attitudes are changing, but even if this particular approach was to be widely used, it has to be remembered that it is not the ultimate answer. It has long been known that micro-organisms evolve and thrive under strange conditions, feeding on whatever organic materials are available, but one has still to find the right micro-organisms and the right conditions for fast feeding – feeding the 'bugs' and destroying the waste. Mineral nutrients, such as nitrogen and phosphorus, are as essential for living creatures as water and, if aerobic bacteria are used, they will require oxygen as well. Too high a concentration of toxic organics may well inhibit bacterial growth and even kill the micro-organisms. Modifying the environment in order to encourage growth can include the deliberate addition of the appropriate nutrients, pH adjustment and oxygen enhancement. Microbes are usually non-pathogenic: that is, they normally die off after their toxic waste (their food) has been degraded and so no longer exists in the form they need. Some companies, notably Polybac, are specializing in what can be called 'tailored' microbes, each designed for a specific function. A typical approach is to examine numerous naturally occurring microbes in the laboratory and selectively breed those seen to be the most effective in any particular case.[9]

We have already mentioned that the cost of biological treatment is likely to be less than the more usual forms of treatment. A demonstration plant at a hazardous waste dump site, named French Ltd, located near Houston, Texas, has confirmed that the cost of biological clean-up is about one-third of the conventional techniques, whilst it takes less than half the time. This particular site was a licensed disposal site during the 1960s and had a 7-acre lagoon. It operated for about 10 years, until it was closed by the State in 1971. The site took waste from perhaps a dozen companies, and the lagoon carried more than 80,000 cubic yards of sludges and there was some 70,000 cubic yards of contaminated soil by the time it was closed. The contaminants included PCBs, chlorinated hydrocarbons, phenolics, pesticides and metal ions ranging all the way from arsenic to zinc. Research by a Houston-based company,

Resource Engineering, helped to establish the optimum nutrient requirements for bacteriological treatment.[10] A small part of the lagoon area was isolated by a bulkhead and air spargers installed. Analysis of the water and sludge over a six-month period indicated that the level of bacterial activity increased by several thousand fold, whilst there was an increase in chloride ion concentration. This was a clear indication that the micro-organisms were successfully breaking up the organic chlorine wastes into simpler compounds. Based on these results, it was estimated that a complete clean-up of the site could be achieved in about three years, at a cost of around US$50 million. On the other hand, clean-up by incineration was estimated to take up to eight years, at a cost of some US$135 million. So the biological approach is both technically feasible and commercially viable: one wonders why there is such a lack of enthusiasm.

For underground contamination as well, such as that caused by leaking storage tanks, the biological method is cheaper than any other *in situ* method. Sometimes it is the only feasible method of tackling the problem. In the case of industrial waste water, specially adapted microbes can help reduce the biological oxygen demand levels to within the stipulated limits, so that it can be discharged in municipal sewers and open water courses. But to have real confidence in the approach, much more needs to be known about the way microbes work, how they can be adapted and adapt themselves to specific situations, and how the environment can be modified to increase microbial activity and efficiency of operation.

From toxic waste to harmless byproducts

The chemical and process industry is by far the largest contributor – perhaps to the extent of some 70 per cent – to the toxic and hazardous waste that has to be handled. Depending upon its origin, each specific effluent will contain perhaps one, or a limited spectrum of toxic components. The main sources and routes for the discharge of such wastes are:[11]

- deliberate release, as in the seas, hoping for rapid dispersal and biodegradation of the toxic elements by marine micro-organisms;
- Mining leachates discharged into rivers, lakes and ponds either via surface run off or ground water;

- Accidental spillage, although this is rare;
- Deliberate release of chemicals, such as fertilizers, pesticides, public health biocides and paints containing triethyl and tributylin.

Accidental spillage will often involve the release of very toxic chemicals and can result in severe and unpredictable effects. The deliberate release for bona fide purposes, the last of the possibilities defined above, whilst it is done to nourish crops, kill pests, diseases or weeds, or protect steel structures, will inevitably lead to the release of the toxic chemicals concerned into the environment, even when carefully and properly handled, but nothing can really be done about this in terms of treatment, whether biological or otherwise. However, release in the seas and waterways can, and should, be dealt with. A host of organic chemicals commonly present in such effluents can be profitably removed or reduced by biotechnological processes.[12] It seems that biological treatment processes require a comparatively small capital investment and have low energy requirements. They are environmentally safe and remain effective at low concentrations. This can be crucial where very dangerous wastes in minute concentrations are involved. What is more, the process does not generate waste, and the microbes are strong and self-sustaining. Biological systems, such as activated sludge treatment plants and trickling filters, reduce the biochemical oxygen demand as well as the chemical oxygen demand, take out the suspended matter and can also deal with xenobiotics and other recalcitrant organic effluents. For these, a mixed culture has proved the best, since co-metabolism is often a vital factor. Mixtures of bacteria are already being marketed for such purposes.

In addition to bacteria, enzymes can also play an important role in removing specific compounds or a specific xenobiotic from wastewater containing a mixture of toxic chemicals. However, unless stabilized by immobilization, their active life is rather short. There are, however, some specific applications of enzymes already in use. These include:

- peroxidase from plants, used to eliminate phenols and amines;
- cyanide hydrase from fungi, used to convert cyanide into foramide.

Imperial Chemical Industries (ICI) have pioneered the development of the latter under the product name 'Cyclear'. This is now on the market as a second-generation product said to be less sensitive to fouling from heavy metals. It seems that the potential is there, but is it being properly utilized?

Engineered organisms

The engineering of organisms for specific purposes is suspect: the general public has a real fear of what is sometimes called 'genetic engineering'.[13] Microbes are generally specific to each toxic chemical and each application, but if none of the naturally occurring microbes are suitable for any specific case, then they can be specifically engineered. This is certainly a more scientific and efficient approach, the only trouble being that these 'tailor-made' microbes may themselves be a hazard. One safeguard against this is provided, at least in the United States, by the regulatory control of the EPA, which will not permit the use of any such hazardous bacteria for fear of creating a still bigger problem than the one the bacteria are designed to solve.

Study of the ISI Press Digest lets us see that this is now a 'hot topic', with a range of articles under themes such as 'bacteria for a better environment: microbes take on messy jobs' and 'bacteria might control their own evolution'. The titles of the articles reviewed give a good impression of the range and scope of present developments in this fascinating field:

- 'Anaerobes for improved sewage treatment'[14]
- 'Eating away oil spills'[15]
- 'Cleaning up a chemical dump'[16]
- 'Genetic engineering to increase potential power'[17]
- 'Cells may "choose" mutations'[18]
- 'A genetic switch in time saves line'[19]
- 'Randomness doctrine never properly tested?'[20]

However, recent work by the OTA should allay the fears that surface when this subject is discussed.[21] The field testing of genetically engineered organisms should mean that potential problems can be anticipated and prevented. Small-scale field tests, such as are proposed over the next few years, are unlikely to pose any uncontrollable environmental problems and the tests themselves should help resolve uncertainties that would otherwise arise in the commercial-scale use of genetically engineered organisms. Such small-scale field tests are considered to be a far more reliable guide than experiments in greenhouses or other safe, but unreal, environments.[22]

Genetically engineered micro-organisms hold promise in many more areas than waste management: new vaccines, microbes that serve as pesticides or fertilizers are some of the many possibilities. The potential is such that it is expected that this application of the biological approach to numerous human activities could eventually prove to have far less impact on the environment than the present conventional technologies.

Continuing caution is, however, to be commended, since there is potential risk when almost any organism is deliberately introduced into a new environment. There is always the possibility of undesirable consequences which can never be precisely predicted because even the simplest of ecosystems is very complicated. However, the work now being carried out by the OTA should allay fears. The OTA is a non-partisan analytical agency of the US Congress, that helps Congress deal with the highly technical issues that increasingly confront modern society. The potential for the use of microbes in the field of hazardous wastes is our immediate concern, and Dr Alan Goldhammer of the Industrial Biotechnology Association sums the present situation up as follows:

> We do have [microbial] strains that have been manipulated by traditional technologies which can degrade polychlorinated biphenyls and a number of chlorinated hydrocarbons and aromatic compounds ... a number of companies ... have demonstrations being run ... at various inactive waste sites using microbial treatment technologies for cleaning up waste.[23]

This biological chemist goes on to warn that when a waste carries a diverse variety of contaminants, whilst the micro-organism may degrade one contaminant, another contaminant may kill it.

What does the future hold?

It seems that biotechnology is full of future promise. There are many exciting developments in prospect, such as genetically engineered microbes capable of consuming the more difficult wastes that are now coming from certain industrial processes. Some possibilities are already apparent, but they await commercial exploitation, pending the resolution of the ethical and political problems associated with the general release and use of genetically engineered organisms. However, there are many natural and naturally-adapted micro-organisms that are not restricted in their use as are the genetically engineered organisms, and work is proceeding apace with these. Adapted bacteria are already available for treating refinery, pulp and paper plant waste, previously the most intractable of volume wastes for safe disposal. It seems the 'bugs' are already busy, battling to clean up hazardous wastes.

Apart from genetically engineered micro-organisms, there is also the possibility of manipulating aromatic-degrading micro-organisms. At the moment, this is somewhat speculative, but by no means an impossible dream. The best researched of the biochemical processes is probably the activated sludge system but, even so, the development of a proper model for this process still presents a challenge to the ingenuity of scientists and technologists.[24] Another environmental problem, that of odour, also awaits solution through biological processes. If the composition of odours were more predictable and less variable, a rational design procedure might well be developed, but at the moment little can be done. There is also scope for the development of biological formulations that are tolerant to an acidic environment: this could make a substantial contribution to the rehabilitation of water and soils contaminated by acid rain. Of course, the best alternative is to remove the pollutants at source, as has been emphasized here throughout, and perhaps the biotechnologist may eventually provide a biological solution even to this problem! What is very clear is that environmental biotechnology embraces a wide range of disciplines, the relevant disciplines varying with the problem that is being handled. A biotechnologist must remember, therefore, that he cannot ignore any discipline: it may well have something to offer. The converse is, of course, also true: no discipline should ignore environmental problems.

References

1 Vesiland, P.A., Pierce, J.J. and Weiner, R.F., *Environmental Engineering*, (2nd edn), Wiley, New York, 1988.

2 Gaudy, A.F. Jr., *Elements of Bioenvironmental Engineering*, Engineering Press, USA, 1988.

3 Winkler, M., *Biological Treatment of Waste Water*, Ellis Horwood, UK, 1981.

4 Foster, C.F. and Wase, D.A.J. (eds), *Environmental Biotechnology,* Ellis Horwood, UK, 1987.
5 Goldstein, N. '1988 Biocycle survey – Steady growth for sludge composting', *Biocycle*, **29**, November–December 1988, pp. 27+ (97 pp.)
6 Bertoldi, M.D. *et al.* (eds), *Compost Production, Quality and Uses*, Elsevier, London, 1987.
7 Schell, J.E., *Earthworm Ecology*, Chapman & Hall, New York, 1983.
8 Verbanic, C., 'Bugs and the hazardous waste battle', *Chemical Business*, January, 1988, pp. 18+ (4 pp.).
9 'Waste disposal – garbology', *Economist*, **306**, 13 February 1988, p. 82.
10 Ibid.
11 Mabbett, T., 'Destroying poisonous waste', *Review*, May 1988, pp. 56+ (2 pp.).
12 Knowles, J. and Wyatt, J.M., 'The potential for biotransforming toxic wastes to harmless byproducts', paper presented at the Biotech 87 Conference, London, May 1987. Proceedings published by Online Publications, London and New York.
13 Stallworthy, E.A., 'Genetic engineering: growing pains', *The Chemical Engineer*, (**451**), August 1988, p. 39.
14 Brody, J.E., 'Anaerobes for improved sewage treatment', *New York Times*, 3 November 1987, pp. Cl, C4.
15 Strauss, S., 'Eating away oil spills', *Technology Review*, April 1988, pp. 12–14.
16 'Cleaning up a chemical dump', *Economist*, **306**, 13 February, p. 88.
17 Rojo, F., 'Genetic engineering to increase potential power', *Science*, **238**, 4 December 1987, pp. 1395–8.
18 Brody, J.E. 'Cells may "choose" mutations', *New York Times*, 13 September 1988, p. C11.
19 Rensberger, B., 'A genetic switch in time saves line', *Philadelphia Enquirer*, 9 September 1988, p. 9D.
20 Cairns, J. *et al.*, 'Randomness doctrine never properly tested?', *Nature*, **335**, 8 September 1988, pp. 142-5.
21 'Field testing engineered organisms', OTA Report, Genetic and Ecological Issues, US Government Printing Office, Washington DC, 1987; 'Field testing engineered organisms', *Chemocology*, (summary article) Chemical Manufacturers' Association, Washington, DC, June 1988, pp. 6–7.
22 'Clearing the air on genetic pollution', *Economist*, **307**, 28 May 1988, p. 85.
23 'Field testing engineered organisms', *Chemocology*, op. cit.
24 Foster and Vase, op. cit.

Chapter 17

ONLY ONE EARTH

We now look at the environmental issues involved in waste management. These are vital issues that have been largely ignored until recently, with the result that mankind now finds itself confronted with a number of very serious problems relating to the environment. We have already seen that not only the land but the seas are in trouble. Coastal waters, especially, are highly polluted, contaminated with toxic industrial waste and raw sewage. Scientists are also concerned about the spate of algae blooms, the so-called 'red tides', which have been linked to the nutrient run-off from the use of fertilizers on land.[1] A clean-up demands a global approach, for it is indeed 'one earth'.

Numerous books, monographs and journals deal exclusively with environmental issues and some of these will be referred to as we come to aspects of the subject with which they deal. The six-volume series entitled *Wastes in the Ocean*, published by John Wiley of New York is part of a total of some 60 books by this publisher relating to environmental science and technology. This indicates the breadth of treatment, and of course Wiley are not alone in publishing books dealing with such issues. Typical of the many journals devoted entirely to this subject is *Environmental Management*, published by Springer International. Started in 1977, and described on the cover page as 'an international journal for decision makers and scientists', it is published six times a year. It was hailed by its publishers as a:

> ...unique journal which presents complementary and contradictory ideas ... to pinpoint and properly assess environmental problems and examine interdisciplinary solutions ... [it] focusses on real problems by providing a forum for the discussion of ideas, findings and methods

> that have been, and can be applied to individual environmental management programs ... covering a broad spectrum of conservation, preservation, reclamation and utilisation, the journal publishes material dealing with ecological modelling, resource management, energy, hazard response, environmental monitoring, and hazardous substances.

That is certainly an all-embracing statement, and we have already dealt with many of the aspects of waste management they mention. However, if the various subjects we have dealt with so far are reviewed, it should be realized that they largely deal with methods of treatment and disposal, which in their turn create their own problems – problems which, more often than not, are left for posterity to deal with. The only absolutely sound solution, as we have also seen, is to design and utilize everything that we need and use such that there is no waste. As we have pointed out, waste is in fact created at the design stage, so that is the stage at which its formation should be prevented. But that is an ideal and meanwhile we have to face the fact that we are left with mountains of waste that require treatment or disposal, with disposal rather than treatment the most usual choice.

What is encouraging is that there is an ever-growing interest and concern about the worsening environment. The *National Geographic Magazine* asks: 'Can man save this fragile earth', and publishes nine articles which all in effect call for a new era of global responsibility.[2] This particular issue of the magazine has a dramatic cover page, which the editorial refers to as follows:

> On our holographic cover, an elegant crystal earth shatters after being hit by a bullet. True, human destruction of the real planet moves slower than a bullet, but unless we can change our ways, the result will be just as shattering.

It is their hope that their cover and the articles may inspire just one leader to create a 'giant leap in the environment'. It is indeed true that politicians are now taking an increasing interest in environmental concerns, as is illustrated by an article with the title: 'Changing colour – Green is the world's new political colour'.[3] We are told:

- Pollution is an issue in America's presidential campaign for the first time ever.
- Some of Mikhail Gorbachev's recent speeches have put almost as much stress on ecologia as on perestroika.

- Deng Xiaoping [has declared that] China's burgeoning industry is creating too much smog.
- Margaret Thatcher has begun worrying about the ozone layer.

Yet another article plays on the word 'lime': the green additive to fruit drinks.[4] But lime is also the chemical used to neutralize the acid in lakes. It is said that whilst making the air cleaner is chemically easy, it is politically and economically difficult. The fear is that today's solutions may well be tomorrow's problems. For instance, although lime can neutralize an acid stream or lake, it puts a lot of calcium into the water and no one knows what effect this might have.

Our ailing seas

Seeing that disposal has been, more often than not, the preferred option, our entire environment – air, land and sea – is ailing. Let us begin by considering the last in that list, the sea, although, being more immediate and familiar for most of us, it is the first two that come more immediately to mind when we start to discuss our environment. Nevertheless the seas and their condition are crucial: the oceans cover some 70 per cent of the earth's surface and are estimated to contain some 1.2 billion cubic kilometres of water. Because of its apparent inexhaustability as a reservoir, the impact of waste disposal on the oceans has been a much neglected area, but the oceans play a very important role in our general well-being in many ways. For instance, millions of people worldwide depend on the oceans for their livelihood and for recreation. In the United States, for example, some 75 per cent of the entire population is expected to be concentrated within 50 miles of the shore (including the Great Lakes) by 1990, as a result of a steady trend in that direction. In addition, the oceans play a major role in determining and controlling the weather, whilst coastal waters play a most important role in the life chain. Yet it is the coastal waters that are by far the most polluted. A feature article on this subject declares: 'the coastal waters of the world are under terrifying assault from garbage, toxic chemicals, heavy metals and human waste.'[5]

Despite the research that is taking place, the long-term effects of pollution on coastal waters is not known, but if the food chain which starts in coastal waters is broken, the impact could eventually be catastrophic. Then, apart from the health risks associated with coastal pollution, the financial impact can be enormous in terms of loss of sea catch and lost revenue when beaches are closed. Seafood is growing in importance as an element in people's diet, and polluted waters are certainly not going to serve that end. Industrial activity is also growing steadily, with the result that some areas, such as the Baltic and Tokyo Bay are under great strain, ever-increasing quantities of toxic chemicals being spewed out into these seas. It is suggested that the deepwater seas and oceans are in good shape, but are they? The fact is that we do not yet fully understand the complexities of this subject. It may just be a function of time, and that later these areas too will show signs of stress and pollution.

Wherever we look there are mounting problems. Three rivers in Europe, the Rhine, the Meuse and the Elbe, are said to discharge yearly 38 million tonnes of zinc, 14,000 tonnes of lead, 6,000 tonnes of copper and varying amounts of cadmium, mercury and even radioactive wastes into the sea. Ships are another major factor in pollution of the sea: it is suggested that they dump 145 million tonnes of ordinary garbage into the sea every year. One result of this, it seems, is that a host of fish species, such as salmon, oysters, ray and haddock, once abundant in the North Sea, have all but disappeared. The few survivors show signs of skin infections, skeletal deformities and tumours. Then, all of a sudden, in mid-1988, some 1,200 seals out of a total population of some 5,000 on the coasts of the North Sea perished. The reason is thought to be a virus, but why? To what extent has the level of pollution influenced the situation? No one knows. It seems to have started along the coast of Denmark, which is relatively clean, but the environmental campaigners continue to maintain that pollution has played a part, albeit an indirect one. It has been suggested that the seal's immune system may have been weakened by the heavy metals or other toxins present in the water.

Algae blooms are another problem. Usually they are red, and some that have appeared along the coasts of Scandinavia during 1988 have exhibited some strange, toxic properties. These 'red tides' seem to reproduce at a phenomenal rate and stretch as blankets along mile after mile of coast. Their rapid growth has been attributed to the increased quantities of nitrates and phosphates present in the water due to pollution, but this reasoning is suspect because similar 'red tides' have been seen on the coasts of Alaska and at Baja, California, areas where the adjacent land areas are neither heavily populated nor industrialized. Whilst the phenomenon is not new, its extent is. Japan is also suffering severely, and there it is having a serious effect on their sea-farming industry, which is still largely small-scale and carried on by family businesses. It seems obvious that man's growing impact on nature has intensified the problem, but no one knows how.

Whilst the technology and the expertise is most certainly available to ensure that the present polluting processes are largely eliminated, the cost would be very high. However, it may well be that failure to invest now may prove even more expensive in the long run, with much higher investment having to be made later.

Silent Spring revisited

This warning of imminent disaster is nothing new. Nearly 30 years ago Rachel Carson's classic work *Silent Spring* served an extremely useful purpose, presenting in dramatic form the dangers that could, and would, come from the overuse and misuse of the persistent organochlorine insecticides.[6] The very subtle but strong title of her book, implying the complete extinction of all wildlife as a result of the indiscriminate use of insecticides and pesticides was designed to shock mankind into action. To some extent it succeeded. This at least was the conclusion of a follow-up book that appeared some 25 years later, with a similar and very apt title *Silent Spring Revisited*.[7] During Carson's time, the environment was indeed being shamelessly assaulted by a society which aimed for total control of agriculture. Nature is not perhaps as self-cleansing as Carson assumed, but many of her predictions regarding environmental deterioration, the impact on human health, water contamination and waste sites have indeed come true.

It all started with what we might call a 'breakthrough' in the field of agriculture, epitomized by the fact that whilst in the nineteenth century perhaps 90 per cent of the population in the United States worked on the land, now it is a mere 4 per cent. This is due mainly to the extensive use of chemical fertilizers and other inputs, together with mechanization and the use of better farming techniques. Whilst there have been obvious benefits accruing from these developments, at the same time there have been hidden and unknown costs. The hazards for wildlife have much increased, and there are ever-growing toxic effects on humans. It might be thought that the results are almost wholly beneficial: whilst some species in the animal kingdom have suffered, mankind seems to be doing very well, experiencing a general improvement in health and an increasing life-span. Further, thanks perhaps to the warning sounded by Carson, society has responded by restricting and carefully controlling the use of chemicals on the land and the birds still sing! But is it enough?

Carson's warning has led to the development of new technologies and new chemicals, which are now assessed in terms of both the risks and the benefits they bring, whereas earlier only the benefits were assessed. Indeed, it can be said that today the predominant concern is the risks involved: the benefits take second place. Unfortunately, toxicity evaluation is a complex and uncertain process. The answers are not, nor can be, as precise as the actual analytical measurement techniques that are now available. Intuitive value judgements are required and this has led to serious controversy amongst scientists and the legislators. On the other hand, the public blithely assumes that all risk is avoidable and that an abundant food supply is assured. It has to be recognized that some chemicals are dangerous, yet it is said that their use is unavoidable if an abundant supply of food is to be maintained for the general well-being of the populace. But is this really true, or is there a better road? Some biological controls have proved very useful, and combining these with a reduced usage of chemicals may offer an optimum solution. One illustration of this is what is called integrated pest management (IPM). It seems that virtually all the issues raised by Carson are in some stage of correction, and an informed public opinion is emerging. This latter is likely to be crucial to any ultimate major improvement: only the pressure of public opinion seems to have real influence. The options for survival may be many, but they are usually complex and are certainly costly – some are very costly.

Saving our skins

We have seen what is going on in the sea and on land: what about the air we breathe? We have mentioned earlier the ozone problem which seems to be caused by some of the chlorine compounds being released into the atmosphere. The environmental consequences of this appear to be serious enough to warrant a book with the eyecatching title *Saving our Skins*.[8] This study carries the subtitle 'Technical potential and policies for the elimination of ozone-depleting compounds' and is prefaced by a statement by the National Academy of Sciences:

> ...the development of an oxygen-rich atmosphere, with its ozone layer, was a precondition to the development of multicelled plants and animals, and all life forms on land have evolved under this shield. Therefore a large burden of proof is required of those who say that the composition of the atmosphere can be changed with impunity.

The culprit which is believed to be changing the atmosphere is the chlorofluorocarbons (CFCs), which are widely used as a refrigerant, as aerosol propellants, as a solvent, and in foam packing and insulation materials. The CFCs are non-toxic to their immediate users in most circumstances, but their indiscriminate and worldwide release threatens to deplete the upper atmosphere of ozone, the substance that shields the world from potentially harmful ultraviolet radiation. It is said that, if CFCs continue to be used at their present rate, ozone levels may fall dramatically, resulting in both humans and animals suffering blistering sunburn even with an hour of two of exposure to the sun. It is also thought that entire ecosystems, such as the coral reefs, may well be wiped out. The Montreal Protocol, agreed in September 1987, calls for a 50 per cent reduction in CFC production by the year 2000. Since then there has been further evidence of ozone depletion that is causing great concern. At a United Nations Environment Programme Conference held at The Hague in October 1988, the government and industry representatives agreed that the Montreal Protocol should be tightened up to require an 85 per cent reduction in CFC production by the year 2000.[9]

Two other chemicals are similarly suspect, carbon tetrachloride and methyl chloroform, and these also need to be banned. The technology required to recover, recycle and replace these chemicals is already available, but positive policies set up worldwide are necessary for action to be taken. Once the ozone depletion problem was appreciated in the 1970s, some companies did identify some satisfactory substitutes for the CFCs, but these efforts seem to have been later abandoned. The principal reasons were apparently a lack of any positive legislation, and thus no market demand. Whilst some of the major multinationals are taking steps voluntarily to eliminate their use of CFCs, it seems that government regulation is required to ensure really effective action. Various countries, including Britain, are now introducing such regulations, but had these been put in place when the problem first surfaced, economic alternatives might by now be freely available. The high cost of such alternatives is the most immediate problem, especially for the developing world.

Amongst the countries that have taken positive action, Sweden has taken a lead by drawing up a programme for the phasing out of CFCs. The lower usage areas, such as sterilants, spray containers and packaging materials, are to be phased out immediately – that is, before 1990. However, this represents only some 3 per cent of the total usage. The major outlet, rigid foam polyurethane and refrigerants, is to be phased out by 1994–5. The cost of this programme will be substantial, but the cost of not implementing such a programme could well be far higher. It really may be a question of us 'saving our skin'! Recovery and recycling of ozone-destroying compounds must be encouraged by fiscal means, such as heavy taxes on the use of suspect chemicals and incentives for the manufacture and use of safer substitutes. However, to be really effective, it is imperative that the Swedish phase-out programme be adopted amd implemented by all the major producers and users of CFCs and other suspect chlorine compounds. Intensive research and comprehensive international cooperation is really essential in this area, but we doubt whether such a cooperative effort is possible. The Third World, in particular, needs to be assisted in the phasing out of the CFCs and their substitution, but this requires adequate financial and technical assistance from the developed world which created the problem in the first place. Will such help be forthcoming?

Energy, the environment and the economy

Charting growth across the world during the twentieth century can be quite revealing. The population has tripled, gross world production (GWP) has increased twentyfold, whilst there has been a twelvefold increase in the consumption of fossil fuels. The data can be tabulated as shown in Table 17.1.

Table 17.1 Growth of global population, GWP and fuel Consumption, 1900–86

	Population (billion)	Gross World Production (trillion US$ (1980))	Fuel Consumption (billion tonnes coal equivalent)
1900	1.6	0.6	1.0
1950	2.5	2.9	3.0
1986	5.0	13.1	12.0

Mankind may well be on an exponential, suicidal course, since these three indicators are largely responsible for the deteriorating environment. Whilst the need for restraint has occasionally been voiced, cheap energy has fuelled both production and pollution. There has been very little increase in the land area on which crops are grown since 1950, but a dramatic increase in output has nevertheless been achieved, largely through mechanization and the use of artificial fertilizers. But, in the process, the environment has suffered so badly that one must begin to wonder as to the wisdom of such a course. The economist, Herman Daly, paraphrases it thus:

> ... as the economy grows beyond its present physical scale, it may increase costs faster than the benefits and initiate an era of uneconomic growth which impoverishes rather than enriches.

It seems that economic activity may well have reached such a level that further growth in the GWP will cost more than it is worth. Largely due to the burning of fossil fuels, there has been such a build-up of atmospheric carbon dioxide that the earth may be warming up to a degree never seen before. It may well be that farmers will soon be finding that the cost of adjusting to new temperature levels and a changing rainfall pattern is depriving them of the investment capital needed to sustain and expand output. Further, the cost of protecting populations in low-lying areas could well be completely prohibitive. All this poses a dilemma and presents us with a paradox. In improving living standards, we may well be threatening the global economy. It is progress – or is it? We need to redefine progress in the light of the intolerable consequences that now seem likely.

Let us be clear: we are not discussing a possibility but a reality. The carbon dioxide-induced warming, a consequence of the so-called 'greenhouse effect', has now been confirmed by several independent agencies. The greenhouse effect can briefly be explained thus: sunlight heats the planet earth, some of this heat being radiated as infrared energy, some as light reflected by clouds and ice, and some as heat trapped by the clouds. The net effect over many thousands of years has been a fairly stable global temperature. Now, however, carbon dioxide and other gases produced by industry, deforestation and farming are trapping far more infrared energy. This is increasing the earth's temperature by a small but significant amount. It is this that is described as the 'greenhouse effect'.[10] Many greenhouse 'models' predict that the earth will warm by at least 3–5°C over the next century, with perhaps even more significant warming near the poles.[11] The biggest problem seems to be the continuous output of carbon dioxide, but the complex reactions involved remain a challenge to the scientists.[12]

At the end of 1986 a US geological survey team confirmed an increase in temperature of between 2–4°C in the arctic region of Alaska. There is no doubt that air pollution and acid deposition (acid rain) is leading to the destruction of forests all over the world, but most dramatically in the Third World. Satellite data has shown that about 16 per cent of India's forest cover has been lost between 1973 and 1981. The damage is continuing and it is impossible to assess the resulting ecological cost. The overall balance sheet is possibly 'in the red', but its extent is not yet apparent. A changing climate, spreading forest damage and similar ecological disasters are a direct consequence of pursuing economic growth at any price – that is, with a complete disregard as to the consequences. For mere survival, mankind needs to find a way to bring good economic sense and good environmental sense together before it is too late, and the world reaches the point of 'no return' and is beyond redemption. How close is the world to this point of no return? No one knows. No one may know until it is too late!

This very ignorance should surely impel mankind to take drastic action, but it does not seem to. Not only are forests being depleted by acid rain but, more directly, they are being destroyed for fuel or to make way for agriculture. In the tropics, only one tree is being planted for every 10 that are cut down. In Africa the ratio is even worse – one to 29. In the quest to meet the immediate demand for firewood is mankind putting the future of his environment at stake? There are any number of questions in this area for which no one has a sound answer. For instance, how will the extensive deforestation now taking place along the Amazon affect its capacity to recycle rainfall inland? Could it set in motion a self-reinforcing process that would dry out the Amazon region? Would this breach a threshold, so leading to irreversible changes in the climate and vegetation of the Amazon area? The questions are there, but there are no answers at the moment. Perhaps the only answer will be what actually happens – when it happens. The media reaction is well illustrated by the dramatic titles that head articles on the subject. Typical of these:

- 'Endless summer – living with the greenhouse effect'[13]

- 'The global greenhouse finally has leaders sweating'[14]
- 'Ominous signs?'[15]
- 'A dire global forecast'[16]

Stabilizing the chemical cycles

Industry has already altered the earth's chemistry quite drastically, with possible staggering ecological and economic consequences. We have already reviewed three serious and very threatening changes that have occurred. Food security is in question, due to the changing climate; there is a drastic loss of forests due to both pollution and deforestation; there is a growing risk to human health from a polluted environment. All this has reached such a scale that it threatens to disrupt natural systems that have been in motion for many thousands of years. Whilst the situation is as yet uncertain, waiting for a clear picture to emerge may well prove disastrous, since by then it may well be irreversible. Drastic and urgent action is most certainly needed now.

The six basic and most abundant elements – carbon, oxygen, nitrogen, hydrogen, phosphorus, and sulphur – constitute nearly 95 per cent of the mass of all living matter on the earth. Their supply is finite and all life depends upon their efficient recycling through a biochemical process. Indiscriminate and thoughtless industrialization has disrupted this cycle, notably so far as it relates to carbon, nitrogen and sulphur. The combustion of fossil fuels is estimated to release into the system about 5 billion tonnes (one tonne per person!) of carbon per year. Since industrialization took hold, back in the mid-nineteenth century, a total of 200 billion tonnes of carbon is believed to have been added to the biosphere. As a result the carbon dioxide content in the atmosphere is steadily rising, with the related consequence that there is less oxygen free in the system. The carbon dioxide in the atmosphere lets the rays of the sun through, but absorbs the heat radiation emitted by the earth's surface. This 'greenhouse effect' has been mathematically modelled and it has been predicted that the concentration of carbon dioxide in the earth's atmosphere may well double by the middle of the next century. This would lead to an average temperature rise of between 1.5–4.5°C. Whilst this is seemingly a very small change in temperature, it would have a profound effect on the climate worldwide. Similar, though less dramatic, effects are expected to occur in relation to the release of nitrogen and sulphur.

The overall effect of these massive additions to the ecosystem is largely unknown. It seems that the biosphere has a self-regulating mechanism of checks and controls which maintain it in a steady state, much as the human body does, but there must be limits. If excessive disturbance occurs, then, as with the human body, damage must result – damage that may well be irreparable. The self-correcting mechanism may well lose its ability to function properly, with completely unknown consequences. The situation is certainly serious enough to warrant concerted action by all the major countries of the world, but we fear that the initiative may well have to be taken by individual countries, who then go on to encourage the others. Some world organizations such as the UN Economic Commission for Europe and the World Meteorological Organization, are concerned about these issues, but none of these international organizations have any power to impose their will on individual countries. The world as a whole really *must* find the right solutions and implement them!

References

1 'Don't go near the water – the world's polluted oceans', *Newsweek* (cover story), 1 August 1988, pp. 36–43.
2 Series of articles in the *National Geographic Magazine*, **174**, No. 6, December 1988.
3 'Changing colour – Green is the world's new political colour', *Economist*, **309**, 15 October 1988, pp. 15–16.
4 'A twist of lime in a cocktail of troubles', *Economist*, **311**, 27 May 1989, pp. 89–90.
5 Hewitt, W., 'Our ailing oceans', *Newsweek*, 1 August 1988, pp. 46–53.
6 Carson, R., *Silent Spring*, Houghton-Mifflin, 1962.
7 Marco, G.J., Hollingworth, R.M. and Durrham, W., *Silent Spring Revisited*, American Chemical Society, Washington DC, 1987.
8 Makhijani, A., Makhijani, A. and Bickel, A. *Saving our Skins*, Environmental Policy Institute and the Institute for Energy and Environmental Research, USA, 1988.
9 Cookson, C., 'The quest for "ozone friendly" gases', *Financial Times*, 11 November 1988, p. 16.
10 Revkin, A.C., 'Endless summer – living with the greenhouse effect', *Discover* (special report), October 1988, pp. 50–61.
11 'Living in the greenhouse', *Economist*, **310**, 11 March 1989, pp. 89–91.
12 Hileman, B., 'Global warming – Special Report', *Chemical & Engineering News*, 13 March 1989, pp. 25+ (18 pp.)
13 Revkin, op cit.

14 Smith, E.T. *et al*, 'The global greenhouse finally has leaders sweating', *Business Week*, 1 August 1988, pp. 44–6.
15 Ramirez, A., 'Ominous signs?', *Fortune*, **118**, 4 July 1988, pp. 102–7.
16 Begley, S., 'A dire global forecast', *Newsweek*, 11 July 1988, pp. 18–20.

Part V
THE ROLE OF MANAGEMENT

Chapter 18

WASTE MANAGEMENT

Waste is waste, but it need not be 'wasted', nor should it be 'wished away'. It should be managed properly and safely, as we have repeatedly emphasized and demonstrated in the first four parts of this book. Now we come to consider the role of management. Our approach will be partly reflective, partly speculative and even philosophical – especially when we come to consider what the future holds in relation to waste management. Sound and effective waste management is very difficult, yet it is essential for mere survival. The primary problem is that the reponsibility of management in fact extends far beyond the factory fence – at times, even beyond national boundaries. Incidents such as the poisoning of the Rhine following the fire at the Sandoz factory at Basel, the Chernobyl disaster, and the widespread phenomenon of acid rain demonstrate very clearly that toxic contaminants know no boundaries. In this context, international cooperation is a must, and it is good to see that this is happening more and more. Over 500 experts, including government officials, policy-makers, scientists and business leaders attended the first world conference set up to explore the solutions to toxic contaminants in large lakes, and its proceedings have now been published.[1]

However, what the individual householder does, or does not do, is perhaps even more important. What is done with waste created in the home is also very important. The proper disposal of domestic waste demands coordination and collaboration, best organized in the first instance by local communities and authorities, but the collaboration needs to be far wider than that. For effective waste management there must also be cooperation on a national and international scale. There are a number of international agencies, largely UN-related, who have a role to play here, such as the International Labour Organization (ILO), United Nations Environmental Planning (UNEP), the European Commission (EC) and the World Health Organisation (WHO). What is the role of these various international organizations?

International cooperation

The ILO is mainly concerned with the workplace and the workforce, and especially concerns itself with the design and organization of the workplace, so that the work is made easy and the working conditions are comfortable. Subjects like ergonomics feature prominently in their programme, together with the safety and health of the workers. That latter is of course affected by a number of factors, including the handling and treatment of toxic and hazardous wastes, so this aspect of waste management then becomes their concern. Typical of the activities of the ILO in the field is the promotion of seminars and conferences. The Indian Association of Occupational Health sponsored the XII Asian Conference on Occupational Health, held in Bombay on 20 Novemer 1988. The main theme of this particular conference was 'Health and safety at work – the worker's birthright', and Kazutaka Kogi, chief of the Occupational Safety and Health Branch, Working Conditions and Environment Department, ILO, Geneva presented a paper.[2] In this way the organization brings its objectives before an international forum.

The activities of the UNEP programme begin where those of the ILO end. The ILO confines its concerns to the workplace, whilst UNEP concentrates on what happens to the environment as a consequence of any 'misdeeds' at the workplace, such as the discharge of waste and effluents. This is also to some extent the concern of UNECE, whose headquarters are in Paris. Typical of UNECE's activities was the promotion of the first international seminar on the subject of non-waste technology in Paris at the end of 1976.[3] On the other hand, WHO is concerned not merely with the health of individual workers, but also with the health of us all. It is in this way that WHO becomes concerned with the hazards presented by both industrial and municipal waste discharges into the air we breathe and the water we drink. All these are extra-governmental activities, conducted by various UN agencies, but industry is also concerned with both national and international issues. Typical of this type of industrial cooperation is the organization CEFIC (Conseil Européen des Fédérations de l'Industrie

Chimique), based in Brussels. CEFIC have taken a significant initiative with regard to the vital subject of industrial waste and its proper handling, which has been clearly detailed in a 20-page brochure.[4] As we have seen, wastes from the chemical industry can be particularly hazardous, and the chemical industry in Europe, as elsewhere, has accepted its responsibility in this regard. As a consequence, the various industry associations have developed voluntary codes of practice for their members, designed to ensure the safe disposal of waste in accordance with sound management practice. The brochure offers advice in the form of broad principles, since the detail will necessarily depend upon the type of waste, the size of the company and its management style. What is significant is that these principles are being extended outside Europe as well, with the free exchange of information across national boundaries.

Basic principles

The CEFIC approach demonstrates that there are certain principles of general application which are basic to proper waste management. Thus:

- Waste reduction and recovery should be pursued to the maximum extent possible, primarily through process optimization, redesign, economic recovery of residues for feedstock, or for energy production. The exchange of technical and economic information, either directly or through technical or trade associations, should be encouraged.
- Waste management should be complete. That is, even if a contractor is employed, it is the responsibility of the creator of the waste to ensure that the contractor is competent to deal with and dispose of the waste safely, will comply with the relevent regulations and handles the waste in accordance with agreed techniques and procedures.
- Before deciding on an appropriate method of disposal, it is essential to know the composition and characteristics of the waste being handled.
- All relevant authorities, especially the local authority, should be kept fully informed of possible hazards and the corresponding precautions. Customers or processors using the products being manufactured should also be made fully aware of any potential hazard, or risks involved in using the products.

THE ROLE OF MANAGEMENT

Where the size of the company allows, a specialist manager should be appointed to advise and assist production and other departmental heads in the proper management of their wastes. Such a manager can obtain the necessary permits, decide on the best disposal option for specific waste streams, and have overall responsibility for the on-site handling, storage and transportation of the various types of waste that may be generated within the plant. There should be a well-defined procedure for the management of specific waste streams, each having been properly identified and documented. A sort of 'waste manifesto' should be drawn up for each waste right from the time it is first generated to the time it is finally disposed of. Adequate records go a long way towards ensuring proper waste management. With a typical integrated manufacturing facility the volume of this data can be considerable, and a centralized computerized database is desirable in order to keep what is happening under continuous surveillance.

In some cases, waste streams cannot be properly disposed at the site, and it will be necessary to employ contractors who specialize in the transport and treatment of such wastes. The contractor should be carefully selected on the basis of capability and, once again, there must be proper documentation. All the contractual arrangements should be fully formalized by written agreements, which ensure that the waste reaches the final agreed disposal site and that the materials are dealt with in the agreed manner. A manager at plant level should be appointed with the responsibility and the neccessary authority to ensure that the procedures are properly carried out. It is advisable to take nothing for granted; the contractor's facilities and procedures should be inspected at regular intervals. The contract should contain adequate provisions in relation to health and environmental protection, whilst material entrusted to the contractor should be properly packaged and labelled. At least an annual review of the contractor's performance is desirable to identify any deviations from the agreed procedure, so that any necessary corrective action may be taken. Frequent and, preferably, surprise checks on contractor performance are essential, to ensure that the waste is being properly handled in accordance with the provisions of the contract.

The disposal plans must be reviewed periodically in the light of changing conditions, changing processes and technology. A method of disposal once considered appropriate, safe and economical may no longer be so. Not only can there be changing conditions, but also changes and amendments to the regulations. It has to be appreciated that any process change, however small, may result in a significant change in the type of waste generated and the proper treatment thereof. Constant and periodic review will ensure that the procedures being used are optimal in the light of the various options that may be open for use. The site inventory of the types and quantities of waste must obviously be kept fully up-to-date. It is also good policy to keep under review the various available methods of treatment and disposal, with technical descriptions, cost information and the impact of the relevant regulations. Not only the initial arrangements, but all changes that are made, should be submitted to management for general review and approval. There should be site contingency plans, designed to cope with emergencies and accidents, and these should also be reviewed in the light of any changes made in the handling of waste, and the appropriate authorities advised.

The US space programme has the possibility of opening up new ideas in relation to the disposal of waste. This new and emerging field has brought its own problems in relation to the handling and disposal of waste.[5] In this case the word 're-use' is the watchword. It is estimated that without the re-use of the liquid waste and urine, over half the total shuttle payload capacity would be taken up by these wastes, on the basis of a 90-day supply of potable water. The cost/weight ratio for space flight is approximately US$1000 per pound, so that it costs over US$8000 to send just one gallon of water into space. The collection of all waste produced within the space vehicle must function flawlessly and this not only has its problems but presents special design challenges. The problems become ever more acute as the duration of the mission increases. If we apply the principles met with here in a wider context, we have to recognise that the earth is also a closed system. The earth has a finite quantity of water, so recycling becomes an absolute necessity if the system is to be maintained. Perhaps a close and careful study of the problems associated with the shuttle and other space vehicles may advance our knowledge of water recycling on earth quite considerably. It is certainly an aspect worth detailed investigation.

Hidden effects

The indiscriminate and arbitrary disposal of waste by industry, municipal authorities and others over the years has been polluting the environment and, what is worse, many of its effects have been hidden. The capacity of the environment to tolerate such flagrant pollution, whilst vast, is still finite, and in some respects the limit may well have been reached. We have in fact discussed a few instances of this – the impact on the ozone layer, the 'greenhouse effect', where the worst is yet to come – but there are many other examples.

The problem is that, as we have seen, there is often a time lapse of several years, perhaps even a decade, between cause and effect. Processes once assumed to be safe have later proved to have very damaging efects. A renowned Japanese epidemiologist, Professor Masazumi Harada, in a keynote address at the Asian Conference on Occupational Health to which we have already referred, described how waste contaminants can leave a tragic trail of disease and death.[6] Otherwise very fastidious, the Japanese have suffered cruelly from the careless handling and disposal of toxic and hazardous wastes, and Professor Harada has illustrated this with six well-documented studies. The most notorious was the Minamata Bay incident in the 1950s when waste from a caustic soda-chlorine chemical factory, containing mercury, was released into Minamata Bay. The mercury was absorbed by the fish, which were then caught and consumed by thousands of people who then suffered a serious disorder of the central nervous system (loss of motor control), manifesting symptoms of disfiguring and slow death; this became known as Minamata disease, being named after the place where it occurred. Some 2,000 persons are officially known to have suffered from this disease, caused by mercury poisoning. Yet the problem hit central Japan in 1965, more than 10 years after the culprit waste was first discharged into the sea. Although dumping of the toxic waste in the Bay stopped more than 20 years ago, it is feared that dredging of the harbour at the site of the initial disaster in an effort to clean it up could send the mercury levels up once again, bringing it back into the food chain. So waste, whilst 'out of sight' cannot necessarily be 'out of mind': it can make its ugly appearance anywhere, anytime.

Other studies cited by Professor Harada included carbon monoxide poisoning in a coal mine, chronic carbon disulphide poisoning in a viscose plant, and chronic cadmium poisoning, which gives rise to the 'itai, itai', ('ouch, ouch', or 'it hurts, it hurts') disease. This last case was linked in 1961 to rice from paddies polluted by toxic waste from a cadmium mine upriver. It seems the cadmium so softens the bones that they can snap very easily. Many died as a consequence, but the survivors, totally incapacitated, can draw benefits under a unique Japanese law. Under this particular law some 100,000 people who have suffered from air pollution or hazardous waste are said to be drawing benefits. The point we wish to make in relation to waste management, is that *all* waste should be most carefully assessed before disposal, and no chances should be taken. The fact that there is no immediate damage to be seen does not mean that there is no damage: it can surface years – sometimes many years – later.

Waste not, have not?

Thanks to its affluence and the so-called 'disposable economy' the United States has become the world's biggest creator of waste. For instance, fully serviceable cars and refrigerators are thrown away, perfectly good buildings are ruthlessly pulled down to make way for even bigger and higher skyscrapers that are then said to only have an economic life expectancy of some 50 years, half that of those built in the 'good old days'. The popular catchphrase 'waste not, want not' now seems to have been replaced by the phrase 'waste not, have not'.[7] Today no one would trouble to straighten out a bent nail; they would merely throw it away and take a new one. Waste of this sort is not merely tolerated, it has become a way of life. Materials, it seems, are now cheaper than labour, so that what appears to be materially wasteful is now economic necessity and has become a way of life. This approach applies not only at the workplace, but also in the home. The housewife is no longer prepared to scrub her oven clean: rather is it lined with expensive aluminium foil, to be thrown away and replaced. Even the 'self-cleaning' oven is not a solution: it requires energy. Re-usable bottles require more labour in handling. And so it goes on. There has been the proliferation of a disposable economy embracing such divergent materials as plastic cutlery, paper dresses, disposable nappies and hospital supplies such as disposable syringes. But, in the process, an even bigger problem seems to have been created – an enormous volume of waste, some toxic, which must be disposed of. The safe disposal of this waste may eventually cost more than the savings in the first place. Whilst it may be more convenient for the immediate user to use and throw away, that 'convenience' has no value. It contributes nothing, whereas the waste produced has to be dealt with at great cost, ultimately, to society.

It seems that the public is no longer interested in longevity. All it wants is the 'latest model', whether that be a car with electronic controls, a refrigerator with automatic defrosting or a self-cleaning oven. Whatever the reasons for this – it could be the lure of the advertisements and the tall claims they make – an ever growing mountain of waste is accumulating that actually need not be there in the first place. In meeting one need, or solving one problem, a far bigger problem has been created. It seems that yesterday's luxury has become today's necessity, and of course the process preserves many jobs. Indeed, we now have the slogan: 'Buy now – the job you save may be your own'! This is an area of waste management that needs continuing attention, for there is no real need for waste of this type. Products should have a longer life and then be recycled for further use, either in the same, or in another, form. The potential is there: what is needed is public education. Unfortunately, no one today is prepared to pursue such a policy: whilst one can exhort the individual to be less wasteful, pointing out the consequences as we have done, there is unfortunately no real inducement towards action.

Planning: the crux

Natural resources are limited and finite, even though new sources may be discovered as man explores farther and deeper. But the use of the available resources should be planned to ensure their effective deployment and to prevent waste. David Novick, an economist by profession, is convinced that the type of materials planning that prevails when a country is at war is also relevant to the peacetime material crisis.[8] His Program Budgeting, a systematic approach to decision-making, initially introduced into the US Department of Defense, was later extended to all US government departments and agencies. He draws the significant conclusion that scarcities are likely to result not from physical limitations as such, but from a lack of proper planning, poor and unsatisfactory data and the assumption that mathematics, statistics and the computer can be a substitute for good data and good sense as a basis for future policies.[9]

Novick's main premise was that man makes his own problems, and man alone can resolve them. Until 1973 the world economy operated on the premise that the world's natural resources, including energy sources, were unlimited. This outlook spurred economic growth all over the world, but the oil crisis of 1973, followed by another in 1979, brought that to a sudden halt. It became apparent that natural resources were being consumed at a far faster rate than they were being discovered. The rapid pace of applied technology has led to belief that man had achieved complete domination over nature, but the energy crisis was a powerful reminder that this was a mistaken assumption. There are valuable lessons in this.

- Be prepared for the worst: it can happen any time, anywhere.
- Reduce demand by conservation, substitution and new technology.
- Increase supply by exploration, investment and new technology.
- Evolve and implement a proper resource management system.

These are basic issues that have to be faced. The seemingly pessimistic, but realistic, conclusion that has to be drawn is that the so-called 'standard of living', especially in the affluent developed world, has to be reduced. This is a matter of economic necessity, and should be self-imposed for survival. The world economy has to be redesigned. This calls for in-depth and continuous analysis, forethought, planning and time, but it is the only way to deal with the range of critical problems that now confront a wasteful world. The problem is *not* confined to the developed world: the developing world is also wasting its resources and needs to plan its future to avoid the collapse of their emerging economies.

The only real and lasting answer to the problems created by waste is proper management, not only of the waste itself but also of the world's resources. At the moment the world is not short of materials, but it *is* short of ideas and planning. Poor management has been, and still is, at the root of most of the world's problems, not only in relation to waste but in other spheres as well. Management is short-sighted: it fails to take in and assess the wider implications of its actions and it ignores consequences which do not immediately affect its profits. Man is quite capable of reducing material demand when the need arises, but unfortunately this capability only comes into play in times of crisis, such as a war.

It also has to be realized that, with resource planning, there is a considerable time lag between any change of policy and its results. Long-term planning is called for – 10 or 20 years is an appropriate time frame – but such planning is indispensable if the world's resources, in which waste has a vital dimension, are to be managed properly and prudently. So vital has this subject now become that some of the larger multinational companies, such as Du Pont have now appointed a Chief Environmental Officer.

References

1 Schmidtke, N.W. (ed.), *Toxic Contamination in Large Lakes*, **1**, Lewis Publishers, USA, 1988.
2 Kogi, K., 'Application of ergonomics in the field of occupational health', *Proceedings*, XII Asian Conference on Occupational Health, Bombay, 20 November 1988.
3 UNECE Proceedings: *Non-waste Technology and Production*, Pergamon Press, 1978.
4 'Industrial Waste Management – CEFIC approach to the issue', brochure (2nd edn), CEFIC, Brussels, March 1986.
5 Wachinski, A.M., 'Waste management in the US space program', *Journal WPCF*, **60** , No. 10, October 1988, pp. 1790–7.
6 Harada, M., 'Several occupational and environmental poisonings in Japan', Guest lecture, *Proceedings*, XII Asian Conference on Occupational Health, Bombay, 20 November 1988.
7 'Waste not, have not', *Reader's Digest*, February 1967, pp. 33–5.
8 Novick, D., 'Facing up to a world of crisis', *Futurist*, **11**, August 1977, pp. 217–23.
9 Novick, D., *A world of Scarcities: critical issues in public policy*, Halstead Press, John Wiley, New York, 1976.

Chapter 19

FINDING THE SOLUTION

To find a solution it is first necessary to know and understand the problem. Perhaps that seems too obvious a statement to make, but not only is it true, it also is often overlooked. All too often solutions are sought without understanding or stating the problem. A problem well stated is, quite often, capable of bringing one halfway along the road to a solution. It is therefore extremely important not only to determine what the problem really is, but also to state that problem clearly, succinctly and concisely. When properly stated, difficult problems often become simple, thus facilitating a simple solution. Waste management brings with it a number of apparently difficult and very complex problems, but the solution is often so simple that we may well wonder: 'Why did no one ever think of that before?' Part of the difficulty is that thinking tends to get into a 'rut' from which there is no escape: a solution seems impossible and remains impossible. This is often the case with what is the normal logical approach to problem-solving – what is called 'vertical thinking'. What is required is 'lateral thinking'.

Lateral thinking

The *Concise Oxford Dictionary* defines 'lateral thinking' as 'seeking to solve problems by unorthodox or apparently illogical methods'. The two component words, 'lateral' and 'thinking' have long been part of the English language, but the combination, 'lateral thinking' is credited to Edward de Bono by some dictionaries. Lateral thinking is something like creativity, being concerned with the ability to escape from the conventional thinking pattern in order to open up new ways of looking at things, or new ways of doing things.[1] Lateral thinking can often give a new direction by providing a new concept for the solving of a specific and seemingly insoluble problem.

One of us was instrumental in persuading the 'inventor' of lateral thinking to conduct seminars on this subject in India, which attracted an enormous response despite the very high fee that had to be charged for participation.[2] There were no hand-outs or course material, but De Bono expounded his concept of lateral thinking using metres of acetate roll, whilst sitting beside an overhead projector. With a style uniquely his own, he illustrated the concept of lateral thinking as applied to a wide range of real life examples. At the end of three days, some of the participants were describing the seminar as 'the experience of a lifetime'! The solution of a complex problem can, at times, be made simple through the use of a simple drawing or diagram. De Bono is convinced that pictures are more powerful than words, especially for driving home ideas and concepts. He drew simple line sketches as he talked to demonstrate the main points of his argument. Discussing what he calls 'non-verbal sense images for management situations', De Bono says:

> The drawings do not have to be accurate or descriptive, but they do have to be simple enough to lodge in the memory. They should not have to be examined in detail the way a diagram is examined, because they are not diagrams. They are intended to convey the flavour rather than the substance of the situation being described.

We mention and indeed seek to emphasize this concept of 'lateral thinking' since it is essential, if waste is to be effectively managed, that those involved 'get out the rut' so far as their treatment of the problems they have to deal with are concerned. An imaginative approach needs to be strongly encouraged. Elsewhere we ourselves have advocated the need for a simple, innovative solution: for example in our book *Safety in the Chemical Industry*.[3] At the present time the reverse seems to be happening. The response of industry to the demand for safe, environmentally acceptable plants seems to have been to make complex process plants even more complex, by adding more and better hardware, such as highly sensitive gas detectors, controllers and analysers. Unfortunately, it seems that complexity and extreme sophistication in process plant control can be self-defeating. What is really wanted is truly innovative thinking, with a drive towards simplicity, rather than the 'add-on' philosophy which seems to be in favour at the moment. In the case of waste handling and disposal, and the related environmental problems, simple solutions within our understanding and capability are often available, but instead a sophisticated 'high-tech' solution is sought and the problem remains unsolvable. To quote:

> Problems are our new frontier ... we tend to colonize them, greeting each new problem, real or otherwise, as new and precious lands for settlement ... much effort is spent in cultivating and refurbishing the problem so it continues to appear fresh and important.[4]

Simple, basic environmental principles, if properly understood and applied, can help solve seemingly complex problems, as we have already seen in the case of the dreadful PCBs and the damage they have done. The solution is simple: stop using them! But how do you do that?

What actually happened? In the United States the Toxic Substances Control Act of 1976 banned the manufacture, processing, distribution or use of PCBs. The Act also required the EPA administrator to lay down methods for the disposal and adequate marking of products containing PCBs. These were defined as 'biphenyl molecules that have been chlorinated to varying degrees or any combination of substances which contain such substances'. Further, a PCB was defined as any substance containing a PCB at a concentration of 50 ppm or greater.[5] This wording demonstrates the great difference there can be between the intent of a law and the actual rule it promulgates. The term 'PCB' is now associated with the phenomenon where a material that persists in the environment can build up in exposed organisms to the point of chronic toxicity. It is *that* process that needs to be controlled. PCBs are not a single chemical, but a range of chemicals, and not all of them are poisonous. The monochlorobiphenyls were labelled 'culprits' by the law, even though they do not behave like their brethren, the higher chlorinated isomers, but degrade so easily that they never build up in the environment to the extent that they can even be detected. To put both types of PCB in the same category is scientifically absurd. The definition of biphenyls should be changed to include only biphenyl molecules with two or more chlorine atoms substituted for hydrogen atoms. This type of mistake is in fact quite common, with the result that there are many discrepancies and sometimes absurdities waiting to be uncovered in the rules and regulations dealing with toxic chemicals. It is absurdities such as this which, perhaps, have led to the popular saying 'the law is an ass'. The deeper we probe, the greater can become our insight. Then we are better able to make rational decisions.

The use of biological processes is another case where simple solutions can solve complex problems. There is no doubt that many toxic chemicals can be treated safely by the use of voracious but innocuous bacteria. For example, a small company, Detox, located in Texas (USA) has developed microbes that thrive on and eat such dangerous chemicals. Anand Chakraborty, a microbiologist at the University of Illinois, Chicago, has evolved and patented a 'molecular breeding' process that can convert the deadly herbicide 2,4,5-trichlorphenol (another PCB) into harmless carbon dioxide and chlorides. There *are* simple solutions to complex problems just waiting to be discovered.[6]

The zero-infinity dilemma

Toxic and hazardous waste is, by its very nature, risky to handle. Hence there is the need to treat it carefully, handle it properly and dispose of it safely lest it pose a threat to man and to his environment. Unfortunately, risk can never be completely eliminated. As the American humorist Robert Brenchley once pointed out, we cannot escape danger even by staying in bed all day – we could fall out! Risk can be reduced, but only at a cost. The cost of reducing risk to a minimum can be prohibitive; in fact risk can only be completely eliminated at infinite cost. This introduces us to what is called the zero–infinity dilemma (ZID). In a remarkable paper Critchley points out the following characteristics, amongst others, of ZID:[7]

- There is a lack of knowledge as to how an accident could be caused.
- There is a catastrophic loss potential.
- The benefits are not related to the costs.
- The risk, although nearly zero, is indeterminate and its assessment highly subjective.

Risk management, as currently practised, has evolved from the earlier 'safety factor' concept (really an 'ignorance' factor) and now uses sophisticated techniques for quantitative safety and reliability analysis. Risk management contemplates and evaluates the 'maximum credible accident' (MCA) but that is as far as it can go. It cannot cope with the incredible – the ZID. Hence the dilemma.

When it comes to toxic and hazardous waste the same principles apply. The acceptable risk must be as low as a reasonable cost will permit. When the risk is low, the probability of trouble is also low, but unfortunately that probability cannot be eliminated. Then, when something *does* happen, the consequences can be catastrophic, since minute traces of some chemicals can do untold harm – witness the Bhopal tragedy. This is the so-called 'low probability high consequence' (LPHC) event. There are many illustrations to be found in the past: typical of them is perhaps the accident at the Icmesa factory near Seveso in Italy. There the consequences of an extremely small emission of the toxic chemical TCDD became manifest over months rather than days, and have persisted for years. The chemical will not dissolve in water, is extremely persistent and minute quantities can kill.[8] This dilemma does not have a definite answer and what is required is a new approach, both to maintain confidence in the techniques and to ensure that those working in industry, the local authorities and the public are better prepared to meet emergencies that have hitherto been thought impossible. Then, if they do occur – and they will occur, since zero risk is practically unattainable – their consequences can be minimized. There is a continuing need in waste-handling for engineering pragmatism. High technology as such does *not* automatically bring improved processing. What was always required, and what is still required, is sound engineering, using designers who can assess the possibilities and then provide for them – designers, in other words, who are capable of lateral thinking.

Cost–benefit analysis

The desire for absolute safety in the handling of dangerous waste has resulted in the escalating cost of process plants for such purposes, an approach which is effectively self-defeating. This is particularly true in the case of nuclear power plants. The dilemma is equally evident on both sides of the Atlantic:

> Costs are being ratchetted up by the lastest safety requirements. Estimates consistently fail to anticipate the cost of efforts to reduce the hazards of nuclear plants … [and] … putative, intangible reductions in accident possibilities have been introduced which are now being belied by the TMI affair.[9]

The very real dilemma is that what is happening to the nuclear industry today may be happening to industry in general tomorrow. The costs incurred in being 'safer than safe', particularly in relation to waste disposal, are largely invisible. We see the expense of prolonged enquiries, the cost of supporting research, the fees of expert witnesses. The Sizewell PWR Inquiry is a notable example. Government approval was won in March 1987, after a six-year planning inquiry and hearing saga, including 340 days when the inspector sat to receive evidence. Will the plant be built any differently or be any safer because of the inquiry? It may even have proved counterproductive in terms of winning public confidence. Meanwhile, the actual cost of the inquiry includes not only the direct cost, but the far greater cost incurred by the delay in getting the plant commissioned and so producing the 'cheap', safe power which has been the prime objective.

The role of regulation

Regulations and laws are essential to a civilized society to contain evil and protect the common man. Each country has its own system of law and these various systems often differ radically. The differences that exist can lead to much confusion, but they are a demonstration of the fact that law grows and develops to suit the needs, circumstances and national characteristics of each country. This remains true when we come to consider waste treatment, handling and disposal. Yet, in each country, the objective should be the same – to protect the environment and the populace from harm.

Most of the laws relating to the handling and disposal of toxic and hazardous waste are of relatively recent origin. The complexity of the problem – the wide variety of chemicals and diversity of situations – is in fact such that it has proved impossible to have complete regulation for all eventualities. Then, because of the varying backgrounds and differing stages of development, such regulations will differ from country to country. For the sake of simplicity and space, we therefore confine ourselves to describing the situation in the United States, with but a brief reference to other countries. In all probability the United States is the most regulated country in this respect and thus provides us with a most comprehensive example.

Table 19.1 Safety legislation in the United States

Statute Standard	Exposure Area	Agency
Consumer Products Safety Act	Hazardous consumer products	CPSC
Poison Prevention Packing Act	Packaging of hazardous consumer products	CPSC
Hazardous Materials Transportation Act	Transportation of toxic substances	DOT (Materials Transportation Bureau)
Federal Railroad Safety Act	Railroad safety	DOT (Materials Transportation Bureau)
Ports and Waterways Safety Act	Shipment of toxic materials by water	DOT (Coast Guard)
Toxic Substances Control Act	Chemical substances excluding FDA controlled material	EPA
Clean Air Act	Hazardous air pollutants	EPA
Safe Drinking Water Act	Drinking water contamination	EPA
Resource Conservation and Recovery Act	Solid wastes, including hazardous material	EPA
Federal Insecticide, Fungicide and Rodenticide Act	Pesticides	EPA
Federal Food, Drug & Cosmetic Act	Food, drugs and cosmetics, medical devices, additives	FDA
Fair Packaging and Labelling Act	Packaging and labelling	FDA
Public Health Service Act	Biological products	FDA
Federal Hazardous Substance Act	Hazardous household products	CPSC
Occupational Safety and Health Act	Workplace toxic chemicals	OSHA, NIOSH
Long Shoremen and Harbor Workers Act	Shoremen/harbour workers exposures	OSHA, NIOSH

Note: Similar statutes have been put into effect in all Redeveloped countries, but in most developing countries the scope is not so comprehensive.

The key legislation with regard to chemical pollutants and chemical products in the United States is presented in a paper by A.S. West, 'Safety evaluation of chemicals – the regulatory framework'.[10] The list is given in Table 19.1. Of the acts listed, the TSCA (Toxic Substances Control Act) is all-embracing, covering the regulation of both existing and new chemicals. Its implications are far-reaching and it requires industry to furnish the EPA with both technical and business information about production, distribution, use, exposure, health risks and the like in relation to the manufacture or use of the chemical. The results of tests on potentially harmful chemicals have to be made available, with a pre-manufacturing review for new chemicals. TSCA, as the name implies, regulates the manufacture, distribution and disposal of chemicals dangerous to health or the environment, and its authority extends from industrial chemicals through pesticide intermediates to consumer products. When one considers the current, though negative, definition of a chemical contained in the question: 'What is not a chemical?', the all-embracing scope of the TSCA can be realized. It covers almost everything that happens at the workplace, at home, and everywhere else – even in space and at the bottom of the oceans. Of course, the United States is by no means the only country to be busy with preventive legislation. A new waste avoidance and waste management act came into force in West Germany on 1 November 1986. This act gives priority to waste utilization over its disposal, disposal being the last resort. But we suspect a problem remains with enforcement: that must be difficult.

Is regulation the best approach? One realistic alternative is said to be to make management fully responsible. This concept has led to at least one paper on the subject with the unusual and provocative title: 'Exit the safety inspector?'[11] While regulations and a system of inspection exists, the company can feel absolved from responsibility once it has complied with the appropriate regulations and satisfied the requirements of the inspector. Once, however, the company is aware that a mere meeting of the regulations is no insurance or excuse if trouble comes, it is far more likely to ensure that its provisions for waste management are both adequate and being observed not only to the letter, but more particularly in the spirit. There is no doubt that non-regulatory methods of managing waste deserve serious consideration by governments. Self-regulation is certainly receiving serious consideration in many quarters, as is evidenced by a series of papers under the general heading 'In-house self regulation of health and safety' presented at the Annual General Meeting of the Industrial

Health And Safety Group of the Society of Chemical Industry in London in June 1986.[12] The crucial elements of self-regulation are seen to be:

- a commitment by senior management to health and safety;
- the existence of effective technical and motivational control;
- levels of education and training that will permit effective performance of the required roles.

Of these three elements, the commitment of senior management is by far the most important, since this will bring about the other two factors.

The regulatory dilemma

There is no doubt, as we have said, that laws are essential in a civilized society, but do we have too many or too few when we come to consider waste management? Perhaps both is the conclusion of Mendeloff in a most unusual book: 'The dilemma of toxic substances regulations'.[13] The introduction points out that in the author's judgement there has been both overregulation and underregulation. Some in industry argue that the legal standards set are far too strict – in fact so strict that they cannot be met in practice and hence they would be better off without them. Others, such as the environmentalists, maintain that standards are too weak and have been implemented far too slowly, with the result that people have been left unprotected against a wide range of hazards, of which toxic waste disposal is but one. Both sides agree that rules are necessary, but it is also recognized that they can never be completely adequate. Mendeloff recommends a three-part legislative reform process that he feels would resolve the dilemma:

1 Allow a proper balancing of the true costs and the true benefits.
2 Create a two-track rule making system, requiring a lower standard of proof when the cost of conforming to the rule is low.
3 Adopt *en masse* all changes in exposure limits recommended by that expert body ACGIH (American Conference of Government Industrial Hygienists).

This seems to be both a pragmatic and sensible approach. In effect it incorporates the self-policing concept, with random checking by the authorities whose officials can then concentrate on the more vital issues. Those establishments and processes that present a substantial hazard can then be carefully and closely assessed and the pubic safeguarded.

References

1 De Bono, E., *Conficts – a better way of doing things*, Harrap, London, 1985.
2 De Bono, E., 'Lateral thinking for management', seminars organized in March 1987 by the Taj Continuing Education Programme at Bombay at Delhi (coordinator O.P. Kharbanda.)
3 Kharbanda, O.P. and Stallworthy, E.A. 'Choosing the right process' in *Safety in the Chemical Industry*, Heinemann, 1988, ch. 26.
4 Isaacs, J.D., 'Challenges of a wet planet', *Chemtech*, **10**, March 1980, pp. 141–3.
5 Neely, W.B., 'Complex problems – simple solutions', *Chemtech*, **11**, 1981, pp. 249–51.
6 Huisingh, D. and Aberty, J., 'Hazardous wastes – some simple solutions', *Management Review*, **75**, June 1986, p. 46+ (5 pp.).
7 Critchley, O.H., 'A new treatment of Low Probability Events with particular application to nuclear power plant incidents', *Progress in Nuclear Energy*, **18**, 1986, pp. 301–57.
8 Kharbanda, O.P. and Stallworthy, E.A., 'Controlling toxic chemicals' in *Safety in the Chemical Industry*, Heinemann, 1988, ch.11.
9 Komanoff, C., 'Cost evaluation at US nuclear power stations', *Minutes of Evidence*, House of Commons Paper HC 397-V, HMSO, 12 March 1980.
10 West, A.S., 'Safety evaluation of chemicals – the regulatory framework', *Plant/Operations Progress*, **5**, January 1986, pp. 11–16.
11 Baker-Conseil, J., 'Exit the safety inspector?', *Process Engineering*, April 1986, p. 29.
12 'In-house self regulation of health and safety', *Chemistry and Industry*, London, 15 September 1986, pp. 598–606.
13 Mendeloff, J.M., *The dilemma of toxic substances regulations*, MIT Press, USA, 1988.

Chapter 20

FACING UP TO THE FUTURE

Having discussed the various disposal and treatment techniques that are available, the various methods that are being adopted for the proper management of waste, and the related environmental issues, we believe that we have dealt with the more vital aspects so far as they affect the role of management. We now propose to take a look at the shape of things to come, since management also needs to plan ahead. Where do we go from here?

Let's take stock

As we have seen, many alternatives for proper waste management are already available, although some of them are not being used to the extent that they might well be. However, we must point out that there is a lacuna in two respects: there is a lack of technical design information and a lack of objective means for evaluating and comparing the various competing technologies. Much, it seems, has to be taken on trust. When it comes to comparing alternative technologies for the treatment of waste, more often than not we seem to be comparing 'apples' and 'oranges'. To be strictly comparable, the competing and alternative methods of disposal must achieve the same results – that is, they must serve the same or a very similar function. To take an example, while an incineration process destroys organic compounds, landfill merely stores them. Hence the two cannot be properly compared. It may well be true that landfill is cheaper than incineration, but that does not mean that it is the preferred choice, since the end result is very different. If it is agreed that proper waste management includes the elimination of waste wherever possible – and that is a stated goal of the RCRA – then the degree to which elimination is achieved should also be weighed in the balance. It is not enough just to transfer waste from one location or medium – air, land, or water – to another. Good waste management will seek to achieve the maximum possible elimination of waste, not merely its safe disposal. In the ideal case, of course, there should be no waste at all.

It is also clear from our review of the various aspects of waste management that there are many problems that still have to be overcome, even though a wealth of fundamental knowledge is available. This store of knowledge needs to be creatively adapted to specific problems by chemists, engineers, microbiologists, physicists, geologists, toxicologists and others. The implementation of effective processes is being delayed by legal, political and social discord. There is much fruitless and even unnecessary argument. What is most disturbing is that the means are there, but they are not being used. For instance, a case study for the destruction of 7 kg of tetrachlorodibenzodioxins, a most virulent toxic chemical, showed that the technical expertise existed to solve even this most difficult pollution problem.[1] Simulation programmes are an economic and quick way of arriving at a cost-effective technique for the proper management of toxic and hazardous wastes. What is more, such programmes, with the method of application, have been published.[2] What seems to be lacking is not the expertise but the necessary will and commitment. Dare we say that the major reason for this lack of commitment is that there is no money in it? The profit motive prevails and even dominates most business activities.

We have also seen that waste management is a global problem, not only in the sense that it is worldwide, but that the successful treatment of waste involves crossing national boundaries. Fortunately this aspect of waste handling and disposal is now beginning to be tackled systematically. Public awareness exists and is increasing. What is more, this public awareness has, over a period, led to the appropriate legislation and related regulations being put into place in a number of countries. There is a growing realization as to the potential benefits of source control – stopping the trouble before it starts or solving the problem at its 'roots'. This aspect is now being earnestly pursued. Non-waste technology, waste recovery and recycling are always to be preferred over landfill or underground storage. The former approach tackles the hazardous waste problem at its source, whereas the latter only postpones the problem. It may postpone it for a very long time but it is still only postponed, not solved. Even chemical destruction and incineration, apart from being very expensive, can sometimes seriously pollute the environment. On the other hand, recovery and recycling is at times economically attractive and is likely to become even more economic as time passes, and certain basic materials become ever scarcer and more expensive. Dangerous wastes can also be detoxicated by chemical, physical and biological processes, and this approach is now being adopted much more frequently. In the case of the dreaded PCBs, most efforts have been directed towards destruction, but when this is done, whilst a cost is incurred there is no corresponding direct benefit in the form of recovered valuable products. A notable exception in this particular area is the extensive work on the treatment of PCBs being carried out in Japan. A process is being developed to reduce PCBs by hydrogen in the presence of metal catalysts in order to obtain high-quality, useful biphenyl. It seems that this particular approach is being pursued because it is likely to be profitable: how much better if it was being pursued because it was a means of resource recovery – a most desirable goal.

Let's keep politics out of it!

Unfortunately waste management has now become a political issue, instead of remaining the purely scientific and technical issue that it ought to be. There are thousands of toxic chemicals, and there is no such thing as a 'typical waste'. Waste may be harmless in itself, but it can also be corrosive, reactive or toxic. Further, there is a great deal of technical argument about toxicity thresholds: what is 'safe'? However, it seems that it is in fact eventually the general public who decides what it wants and what it will tolerate. Some of the hazardous waste treatment methods that we have described do indeed lead to residues presenting a much reduced risk to the public and the environment – a risk which it seems society is prepared to accept. The concept of regional waste centres illustrates the degree of cooperation that can be secured from the public once it has been demonstrated that the approach is both economic and convenient. For a wide variety of reasons, some of which we have dealt with in our text and briefly summarized above, the preferred technologies for the treatment of hazardous waste have shifted from land disposal to those that achieve the destruction of such waste, with toxicity reduction and elimination. Regulatory limitations have now been imposed that exclude certain low-cost methods for hazardous waste disposal, such as landfill. Waste containing PCBs, solvents and dioxins may no longer be disposed of via landfill, and acceptable alternatives have therefore to be sought.[3]

But what policy should governments pursue in relation to the management of hazardous wastes? The answer to this question is very simple and very brief: none. In fact, that is not the right question. What should be asked is: how can we improve hazardous waste management and so protect public health and the environment? What government policies would help in this respect? Why haven't we done better in this respect in the past? When the past is viewed, we see many failures, but what is their true cause? Is it really lack of regulation? It is certainly true that many of the present problems have their roots in the past, but it seems to us that the main cause was ignorance. There was a lack of adequate public knowledge and understanding and there was limited technical knowledge, so that the impact of the actions taken was not appreciated. If the polity and the marketplace are uninformed, uninformed decisions will inevitably be taken. Fortunately, things have now changed and continue to change. Both governments and the public are now far better informed than they were, and there is a growing concern about the issues involved in waste management, particularly now that it is appreciated that this concerns each one of us *directly*.

Robert Cahn, who wrote the challenging book *Are Pesticides Really Necessary?* has now edited a book with the title *An Environmental Agenda for the Future.*[4] This is a collection of papers where the leaders of 10 major public interest groups propose policy changes and new laws to help stem environmental abuses. Their pronouncements are spelt out in detail and deserve close scrutiny, particularly by the policy-makers. Their proposals could also be a possible basis for a constructive dialogue with industry so that acceptable solutions can be found. Industry and the public should listen to the environmentalists, not try to ignore them as a nuisance. On the other hand, the environmentalist may well have much to learn from those who feed, house, transport and otherwise supply the world with life's necessities. Unfortunately there are no chemists or chemical engineers included in the 10 authors who are putting forward their solutions, which therefore lack the scientific and technical dimension so necessary for a proper consideration of this subject.

Another futuristic study, by Ralph Atkins, examined the problem of domestic rubbish.[5] Despite having been pioneers in the industrial revolution, Britain has no coherent and consistent waste management policy. As we have seen, there is no shortage of ideas; recycling is advocated and the relevant technology is well known and readily applicable. What seems to be lacking is the ability to persuade people, industry and the authorities that domestic rubbish can be an extremely valuable asset, as some communities, particularly in Japan, have demonstrated. What is more, not only can it be a source of valuable products, but its proper handling will help conserve the environment as well. It seems that, to put this into practice demands a radical change in attitudes, which is just not forthcoming. For its processing to be economically viable, domestic waste needs to be pre-sorted at source. This demands both the cooperation of the householder and organization by the local authority. Both are lacking: the will is not there. Meanwhile, the disposal problem grows by the day. Thus, although much is now being done in relation to the proper management of waste, there is still much more to do. The future in this field is full of challenge.

Solving the problem

Whilst a complex set of problems continues to haunt society, at the same time new solutions for the handling of the waste byproducts of the continuously developing society in which we live keep emerging. Policy decisions in relation to waste management have come about more by default than by design. There has been little or no planning. However, more recently attitudes have changed, with the result that the precious air, land and water resources are no longer considered to be usable commodities. Research and development and the extensive exchange of knowledge through international conferences and seminars have helped to provide suitable, sound answers to many of the key questions in relation to proper waste management. There is, of course, not only the what, the how, and the where, but that ultimate and most crucial question of all: who pays? It seems that this present industrialized society has reached a turning-point, or perhaps a point of crisis. At last the realization has dawned that the future is *not* endless – an open, empty frontier. The world's population is growing rapidly, shortages in certain vital materials are imminent and those in a position to take action must foresee and account for the effects of their present policies. A difficult, complex blend of political and economic issues currently exists, and the position is further confounded by legal, chemical and biological problems that seem to defy resolution. A very positive approach is required, but who has the will, the courage or the authority to apply it?

On the political front, the battle is heating up. On the one hand, the public demands that the problem be solved, no matter what the cost, without realizing that ultimately that cost will have to be borne by them. On the other hand, industry strongly resists the sweeping changes that would be the inevitable result. In general the administration sides with industry, fearing the economic consequences. It would be fine if everyone and every country fell into line and acted in concert, but meanwhile those who take steps to deal with these problems incur additional costs which place them at a disadvantage in this competitive world. The painful lessons that come from the careless disposal of waste have not been learnt. The conflict between industry and the public seems not only unresolved, but on the increase. It is a tragedy that narrow political and commercial interest has outweighed the growing public concern.

Toxic and hazardous waste, in particular, has already created much havoc, both in terms of its immediate effects and the overall pollution of the environment, with its even more horrific long-term effects. Whilst the scientists, the technologists and the law-makers are now confronting these problems and are seeking to manage waste in such a way as to make the world a safer and healthier place, it is a slow, painful process. The solution certainly does not lie in the closing down of all industrial activity and reverting to an agrarian existence. With the world population at its present level, and growing, that is just not feasible. That population needs to work, and industrial activity provides it with employment. Industry is here to stay; the work it provides and the products it produces are the base upon which all social benefits and gains have to be built. In addition, however, industrial activity must be organized so that it presents no threat to the environment. *That* is the task of the hour!

But let us be realistic and look at the positive gains. The situation is not as bad as it is sometimes made out to be. In 1900, in the United States, a newborn child had a life expectancy of about 40 years. Now it is over 70 and still rising. A similar situation prevails practically everywhere. As a result, this present generation may well claim to be the first generation that has a reasonable chance of living its life out, rather than having that life cut prematurely short by acute illness. So there are grounds for optimism. This optimism is echoed by the political and social scientist, Aaron Wildavsky:

> The future won't be allowed to make mistakes because the present will use up its excess resources in prevention ... How extraordinary! The richest, longest-lived, best protected, most resourceful civilisation, with the highest degree of insight into its own technology, is on the way to becoming the most frightened.

We can take comfort in that the world is wiser as a result of the hard lessons that have been learnt from the troubles and the problems that have come from the improper treatment and disposal of waste in all its forms, even if there are still many lessons that have not been learnt. At least the past has taught us that sound pollution control, health, safety and proper product management are all elements of effective waste management.

Be prepared for the worst

Based on a nationwide questionnaire in the United States to a wide range of major corporations, a series of conceptual and empirical issues have emerged which may help in the development of a general theory of crisis management.[7] Management should take all the care that it can in the proper management of waste, particularly hazardous waste, yet at the same time it should prepare for the worst – that is, crisis management. Accidents can always happen; then disaster may strike. It is this philosophy which is leading many major companies, including the multinationals, to spend much time, money and effort in keeping the environment clean and free from potentially dangerous chemicals and other toxic materials.[8] Environmental activists and the mass media seem to have been playing a vital and useful role in relation to industry. An industrialist cannot and will not ignore the media, no matter how exaggerated their claims and statements may be. To ignore what is being said to the public is to risk the reputation of the company and lose its market share. Further, when warnings are ignored the company may later find itself involved in time consuming and costly clean-up operations. An adverse public image, resulting from genuine or even exaggerated claims of lack of proper action, will undoubtedly affect its profits. So there is a sound financial motive for taking care.

Yet, despite the best efforts of companies and their managements in respect of waste management, its treatment and disposal, a crisis situation *can* arise and management need to be so organized that such situations can be met. This realization has led to the relatively new concept of 'crisis management', but few companies as yet seem prepared to meet and cope with crisis.[9] As part of overall crisis management there must be a disaster plan in place designed to cope with fires or unexpected pollution. The absence of such a plan, or poor implementation of such plans as have been made, can have a disastrous effect on the company's profits. It is well established that a crisis or a calamity *can* be controlled and managed in order to minimize its effects provided that the appropriate preparations have been made to put a disaster plan into action when disaster strikes.

The environmental audit

Environmental auditing is now seen as an essential part of proper waste management and it has assumed sufficient importance for a full-length book, the *Environmental Auditing Handbook*, to be written on the subject.[10] Written by experts in their respective fields, this handbook provides all the practical information required to institute a waste management audit. Not only will such an approach help in complying with the relevant laws and regulations, but it helps in the identification of environmental risks. The relevant areas are explored from a common vantage-point for industry executives and managers, auditors, engineers, insurance brokers and government inspectors. The entire gamut of environmental concerns, including those arising from waste management, and those affecting the worker's health and safety and the welfare of the local community are dealt with in great detail.

Among the subjects covered are: how to perform an environmental audit; how it can help management; how to minimize the risks associated with hazardous waste, and with air and water pollution; and how some of the major companies perform such an audit. Claimed to be the first handbook of its type, the substantive parts of the book deal with:

- Detailing environmental risks
- Undertaking an environmental audit
- How to protect audit confidentiality

It seems that many major companies now have environmental auditing departments, whilst at least five US states have set up pilot programmes to explore the benefits of environmental auditing. Of course, for maximum benefit the approach needs to be pursued nationwide, even worldwide, but that is a long way away. If pursued vigorously, it would help move government and industry away from their present destructive and costly adversarial system of regulation, with its continual confrontation, towards working together for a common cause.

The public's perception of risk

The disposal of waste involves the taking of risks, but what are the risks involved? Hazardous waste obviously presents special risks, but how significant are they? These are difficult questions and there is no simple, quantitative answer. We have the relatively new discipline of risk assessment, but when numbers are produced in this way we really have very little feel for their significance. We do not know how reliable such figures are, nor what they really mean. The subject of risk and its perception is a very complex matter. Indeed, it seems that what is really significant is the public's perception of a risk, rather than the reality, what ever that may be. It is evident that much of the motivation in relation to the proper disposal of waste comes from the public who perceives a risk and raises an outcry. Public perception of risk is apparently highly subjective and has nothing to do with statistics, nor with the realities of the situation. Man in general must be a most irrational, uninformed and superstitious creature – even stupid. How otherwise can one explain the vast difference between actual risk and perceived risk for some of the technologies, including aviation and nuclear power generation.[11] Most people do not understand probability – one would not expect them to – yet they try to force their unscientific worries on to policy-makers, seeking to influence policy decisions that concern the hazards of modern living. To take an example, experts and lay people were asked to rank the risk of dying in any year from various activities and technologies. The ratings were made with the experts basing their judgement on the known fatality statistics, whereas the public opinion was based on their perception. The results were as follows:

	Expert Opinion	Public Perception
Nuclear power:	1	20
Aviation:	7	12
Non-nuclear-power	18	9
X-rays:	22	7
Surgery:	10	5
Motor vehicles	2	1

Unfortunately, however scientific the approach may be, most people are unwilling to take a purely statistical view of risk. It seems they worry far more about involuntary risks, such as carcinogens in food than about voluntary, avoidable, risks such as rock climbing although, statistically speaking, the risk is far greater with the second than with the first. This may be irrational, but it is true. Thus there is a very wide gap between the real risk and the perceived risk, so far as the general public is concerned. What is more, people react differently, even to the same risk, their reaction depending upon their background and upbringing, since it is this that governs their perception of risk. The rock climber who thrives on risk and enjoys the sport feels safe enough, while the non-climber trembles at the very thought. In pursuit of enjoyment the rock climber simply and deliberately ignores the risk involved.

The lack of consistency in the public perception of risk can be demonstrated by its reaction to the building of an industrial facility. Public opinion polls have been taken that show that it is the hidden fear that is most important.[12] These polls demonstrated that only 25 per cent were prepared to live close to a nuclear waste disposal facility, whereas some 50 per cent were prepared to live close to an insecticide factory. More than 75 per cent, however, were prepared to live close to a natural gas power plant. Yet the scientist would assert that the risk to the public was much greater from the insecticide factory than from the nuclear waste dump. It is clear that ignorance can lead people to believe a risk to be greater than it really is, and this of course results in irrational reactions. Whilst, statistically speaking, it may well be much more dangerous to ride in a car than to fly in an aeroplane, the public perceives the risk differently and considers flying to be more dangerous than driving. Statistics – the numbers game – just does not carry any weight with the public. It is most certainly feelings that count – the perceived risk, not the actual risk.

But what does all this mean in relation to waste disposal and waste management? The public is scared of chemicals, especially toxic chemicals since it is a hidden danger that it cannot assess. Whilst it must be agreed that there are dangers associated with the disposal of toxic chemicals, those dangers can be minimized if proper care is taken. But that is true of everything in life. Whatever we do we shall have to take risks; that is unavoidable. The real question is: is the risk worth taking? Public education is very necessary, to ensure that public pressures are exerted against real risks to the environment rather than assumed risks.

The resources available to minimize risk are limited and must therefore be deployed to the best advantage. The extremes to which the public perception can go is well illustrated by the view expressed by a TV reporter, Tom Vicar, of KCBS TV, Los Angeles, asking whether there is any limit on the toxicity of chemicals, said: 'I'm getting the impression that some compounds are so toxic that they'll kill you even if you think about them' (quoted by Peter Sandman, of Cook College).

We hope that we have demonstrated that waste can be managed safely, without risk to the public or danger to the environment. More can be done, more should be done and we have tried to put the risks involved in perspective, so that both public and private effort towards the proper management of waste can bring the maximum benefit. Now we must leave it to our readers to do what they can.

References

1 Exner, J.H., *Detoxification of Hazardous Waste,* Ann Arbor Science, USA, 1982.

2 Martin, E.J., and Johnson, J.H. Jr. (eds), *Hazardous Waste Management Engineering,* Van Nostrand, New York, 1987.

3 Cahn, R. (ed.), *An Environmental Agenda for the Future,* Island Press, USA, 1985.

4 Ibid.

5 Atkins, R., '2002 an ecological odyssey – laying waste to England's green and pleasant land', *Financial Times,* 14 January 1988.

6 Wildavsky, A., 'No risk is the highest risk of all', *American Scientist,* **67(1)**, January–February 1979, pp. 32–7.

7 Mitroff, I.I., Pauchant, T.C., and Shrivastava, P., 'The structure of man-made organisation crisis'. *Technological Forecasting and Social Change,* **33**, 1988, pp. 83–107.

8 Cotter, M., 'Take good care', *Chief Executive,* March 1988, pp. 32+ (3 pages).

9 Stallworthy, E.A., 'Coping with catastrophe', *The Chemical Engineer,* **(449)**, June 1988, p. 59.

10 Harrison, L.L., (ed.), *Environmental Auditing Handbook – a guide to corporate and environmental risk management,* McGraw Hill, New York, 1984.

11 Allman, W.F., 'Staying alive in the 20th century', *Chemtech,* **18**, December 1988, pp. 720–4.

12 Lindell, M.K. and Earle, T.C., 'How close is close enough? Public perceptions of the risk of industrial facilities', *Risk Analysis,* **3**, December 1983, pp. 245–53.

Author Index

Subject Index

Italic numerals refer to figures or tables